Computer-Generated Phase-Only Holograms for 3D Displays

"Phase-only Fresnel holograms," which can be displayed on a single SLM without the need for lenses or complicated optical accessories, substantially simplifies 3-D holographic display systems.

Exploring essential concepts, theories, and formulations of these phase-only Fresnel holograms, this book provides comprehensive coverage of modern methods for generating such holograms, which pave the way for commercial products such as compact holographic projectors, heads-up displays, and data security enhancement.

Relevant MATLAB codes are provided for readers to implement and evaluate the theories and formulations of different methods, and can be used as a quick start framework for further research and development.

This is a crucial and up-to-date treatment of phase-only Fresnel holograms for students and researchers in electrical and electronic engineering, computer science/ engineering, applied physics, information technology, and multimedia technology, as well as engineers and scientists in industry developing new products on 3-D displays and holographic projection.

Peter Wai Ming Tsang received his Ph.D. degree in electrical and electronic engineering from The University of Hong Kong in 1987. He is currently an Associate Professor with the Department of Electrical Engineering, City University of Hong Kong. Dr. Tsang is actively engaged in research on digital holography and three-dimensional technology, focusing on developing methods for generation, processing, and display of hologram. Apart from research findings reported in international journals, conferences, invited talks, he is also an inventor of 28 patents. Dr. Tsang has served as Guest Editor/ Associate Editor for the *IEEE Transaction on Industrial Informatics*. He is currently an Editor of the *ICT Express* (Elsevier). He also served as Topical Editor of the *Applied Optics*, OSA from 2013 to 2019.

Computer-Generated Phase-Only Holograms for 3D Displays

A MATLAB Approach

PETER WAI MING TSANG
City University of Hong Kong

CAMBRIDGE
UNIVERSITY PRESS

CAMBRIDGE
UNIVERSITY PRESS

University Printing House, Cambridge CB2 8BS, United Kingdom

One Liberty Plaza, 20th Floor, New York, NY 10006, USA

477 Williamstown Road, Port Melbourne, VIC 3207, Australia

314–321, 3rd Floor, Plot 3, Splendor Forum, Jasola District Centre, New Delhi – 110025, India

79 Anson Road, #06–04/06, Singapore 079906

Cambridge University Press is part of the University of Cambridge.

It furthers the University's mission by disseminating knowledge in the pursuit of education, learning, and research at the highest international levels of excellence.

www.cambridge.org
Information on this title: www.cambridge.org/9781108427333
DOI: 10.1017/9781316999172

First published 2021

A catalogue record for this publication is available from the British Library.

Library of Congress Cataloging-in-Publication Data
Names: Tsang, Peter Wai Ming, 1956– author.
Title: Computer-generated phase-only holograms for 3D displays : a Matlab approach / Peter Wai Ming Tsang, City University of Hong Kong.
Description: New York : Cambridge University Press, 2020. | Includes bibliographical references and index.
Identifiers: LCCN 2020026417 (print) | LCCN 2020026418 (ebook) | ISBN 9781108427333 (hardback) | ISBN 9781316999172 (ebook)
Subjects: LCSH: Three-dimensional imaging. | Holography – Data processing.
Classification: LCC TA1560 .T73 2020 (print) | LCC TA1560 (ebook) | DDC 006.6/93–dc23
LC record available at https://lccn.loc.gov/2020026417
LC ebook record available at https://lccn.loc.gov/2020026418

ISBN 978-1-108-42733-3 Hardback

Additional resources for this publication at www.cambridge.org/pwmtsang.

Contents

Preface

According to experts and researchers, holography is likely to be the ultimate solution to 3-D display technology. In simple terms, holography is a method of capturing the full optical field of a 3-D scene in a 2-D complex image. When a hologram is illuminated with a coherent beam, the scene is reconstructed as a visible 3-D image.

With the advancement of computing and display technologies, it is now possible to generate a digital Fresnel hologram of a 3-D object scene numerically. Theoretically, if a high-resolution spatial light modulator (SLM) that is capable of displaying a complex-valued hologram is available, the image represented by the hologram will be reconstructed as a visible 3-D image that can be observed visually or projected onto a 3-D surface. However, in practice, existing SLM devices can only display either the magnitude or the phase component of a hologram, and a pair of SLMs has to be integrated with a precise and complicated optical arrangement in order to display a complex-valued hologram.

An effective solution to the current problem in holographic display is to generate a phase-only hologram instead of a complex hologram, so that it can be displayed with a single SLM. Such an approach will lead to significant simplification in the holographic display system, as well as enhancement of the brightness of the reconstructed image. On the downside, the quality of the hologram and its reconstructed image will be jeopardized. Although research on overcoming this issue began in the 1960s, it was not until recently that computationally efficient methods were developed for generating phase-only holograms at a video rate of over 25 frames per second.

Most of the modern methods for generating phase-only holograms are reported in individual journal articles. Due to the limited space available, the content of these papers is usually presented in a succinct fashion, focusing more on the theory than the practical aspects. This type of scholarly presentation can be adequate for an experienced researcher, but may present difficulties to someone new to the field who needs to acquire the knowledge within a short time frame, or someone who wants to get a full picture of the state of the art (such as an undergraduate student, a postgraduate student, or an engineering practitioner).

This book is intended to provide comprehensive coverage of modern methods for generating digital phase-only Fresnel hologram for 3-D displays – a kind of hologram that enables 3-D images to be displayed on a single SLM without the need for lenses or complicated optical accessories. The research findings related to this technology will have a significant impact, leading to substantial simplification of the 3-D holographic display system, paving the way for commercial products such as compact holographic projectors, heads-up displays, and data security enhancement.

The book is aimed at various readers, including serious researchers, academics, engineers, postgraduate students, and undergraduate students taking courses in digital holography. It includes essential theories and formulations presented in a succinct and readable manner, avoiding material that is less relevant to the main theme. MATLAB code is provided for readers to verify and evaluate the theories and formulations whenever applicable and selected programs can also be found online here: www.cambridge.org/pwmtsang. I hope that this book is a good companion for everyone in the course of conducting impactful research and developing contemporary products on 3-D holographic display systems.

Acknowledgments

I would like to give my deepest thanks to my family for their unfailing support and trust in me. My parents always allowed me to choose my interests and encouraged me to achieve my goals.

I am forever grateful, beyond words, to my wife. She has been by my side since I started my PhD research and my career, and is always positive and encouraging when I take up a new challenge. She has provided me a comfortable and worry-free life so that I can always focus on my interests and work. Her editing assistance and advice were essential to the timely completion of this book.

During the course of writing this book, it has come to me that I have encountered many inspiring and outstanding research scholars. All my friends and research partners (in the order in which I first met them), Professor Ting-Chung Poon, Professor Jung-Ping Liu, Professor Hiroshi Yoshikawa, Professor Taegeun Kim, Professor Changhe Zhou, Professor Hongbo Zhang, Professor Wolfgang Osten, Dr. Kenny Chan, Dr. Xiuling Zhu, Dr. William Wang, Professor Aimin Yan, and Professor Yaping Zhang, have continuously stimulated my work over the decades to make this book possible.

My research in digital holography has also been encouraged by many renowned scholars, experts, and friends, Professor Toyohiko Yatagai, Professor Eun-Soo Kim, Yunlong Sheng, Professor Byoungho Lee, Professor Hoonjong Kang, and Professor Partha Banerjee, who have generously shared their knowledge and ideas. I would like to express heartfelt thanks to everyone I have met on my career path. I am especially grateful to my mentor of 40 years, Dr. F.K. Lam.

I am also grateful to the City University of Hong Kong for providing the research environment and freedom over the past 33 years that allowed my academic endeavors. My deep appreciation goes to Dr. Kayton Cheung, Dr. Wuchao Situ, Dr. Yau-King Lam, Dr. Albert Jiao, and Dr. Yuk-Tat Chow, who have worked shoulder to shoulder with me in exploring the mystical world of holography. Special thanks to Prof. Stella Pang and Prof. Kim Man, who have motivated and supported me to advance my research in a way that I once thought would never occur in my academic life.

Again, I am thankful for all of you reading this book.

1 Introduction to Digital Holography

1.1 Basic Concept of Holography

When light impinges from one medium to another, part of it will be transmitted while the rest will bounce back to the environment or be absorbed into the medium. Figure 1.1(a) shows a ray of monochromatic light (i.e., single frequency or wavelength) L falling on a planar interface between two different homogeneous media M_1 and M_2.The beam is inclined at an angle θ_1 with respect to the normal of the interface. Assuming that M_2 is partially transmissive, and L is a plane wave, the direction of part of the light transmitting through M_2 will be changed from θ_1 to θ_2. This process is known as *refraction*, and the relation between the pair of angles is governed by Snell's law.

If the interface between the two media is smooth, the light that is not refracted will be reflected back as specular (mirror-like) reflection, in which case the angle of incident θ_1 will be identical to the angle of reflection θ_2. For an interface with a rough surface composed of microscopic irregularities, the incident beam will be scattered, resulting in diffusive reflection, as shown in Figure 1.1(b). *Scattering* is a generic term that refers to the change in the propagation path of an electromagnetic wave as it is intercepted by some form of non-uniformities in a medium. Optical scattering results when a beam of light hits one or more tiny particles.

The light impinging on a physical object undergoes one or more of the above-mentioned modes of propagation. Light emitted from the illuminated object surface conveys information on the shape and color of the object. As the emitted light waves fall on the retina of human eyes, an image of the object will be perceived by the brain.

Photography and optical holography are methods for capturing the emitted light waves from an object scene onto a photographic film. When the photographic film carrying the recorded optical signals is presented to an observer, the light waves of the object scene will be partially or fully reproduced to the eyes of an observer. At present, there are two major approaches in capturing the light waves of an illuminated object: photography and holography. Photography is the art of capturing the part of the optical waves corresponding to a two-dimensional (2-D) projected view of an object, while holography captures the entire optical waves. A photograph only presents the intensity or color of a scene; it is a planar image and does not contain any disparity or depth information. With holography, all the light waves emitted by the object are captured,

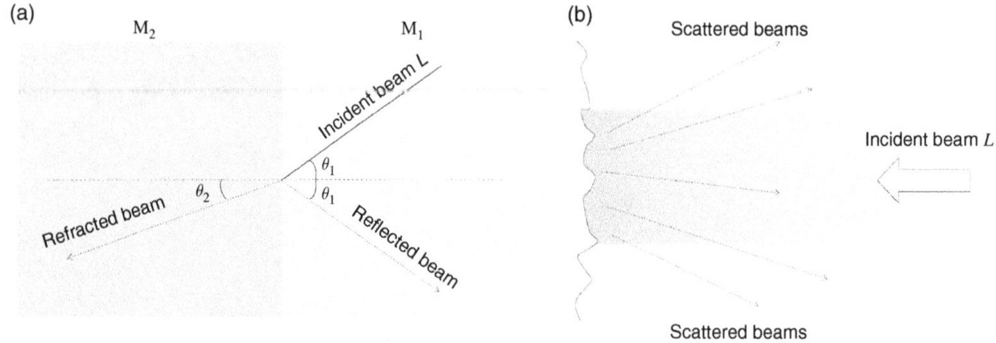

Figure 1.1 (a) Refraction and reflection of a light beam on a smooth surface. (b) Scattering of a light beam on a rough surface.

and hence are capable of reproducing a three-dimensional (3-D) view of the object. These two image-recording techniques are explained as follows.

Light, or optical signal, is a kind of electromagnetic wave in the form of a time-varying sinusoidal wave oscillating with a certain amplitude and frequency. When a person observes a scene, optical waves of different intensities and frequencies will be received by different groups of sensory cells in the retina, creating the impression of color and brightness. The instantaneous amplitude $v(t)$ of a single beam of monochromatic light L is generally expressed as:

$$v(t) = A\cos(\omega t + \theta_L). \tag{1.1}$$

In Eq. (1.1), t represents the time variable, while the terms ω and θ_L denote the frequency and phase of the light beam, respectively. For a monochromatic light wave of constant frequency, the frequency term is sometimes left out, and the representation of the wave can be simplified as $v = A\exp(i\theta_L)$, where i denotes the imaginary unit. In the physical world, an object reflects an infinite number of light beams from its surface, and the collective excitation of these optical signals on the retina leads to the formation of an image of the object to an observer.

Next, we shall explore how the optical waves of an object can be captured. Imagine the light wave of an object, which is sometimes referred to as the object wave, is intercepted by a transparent medium that is capable of recording the amplitude and phase of every optical beam that passes through it, as shown in Figure 1.2(a). For the time being, let's ignore the details of the mechanism behind the recording process. As this medium is illuminated by a coherent plane optical wave (i.e., the amplitude and phase at every part of the wavefront is identical), the light beam that emerges from every point on it will be modulated (in amplitude and phase), by an amount that is identical to that which has been recorded. As illustrated in Figure 1.2, the observer will see the same set of optical waves from either the object or the recording media. Theoretically, the optical waves received by the retina in both cases should be identical, hence casting an impression of the same image.

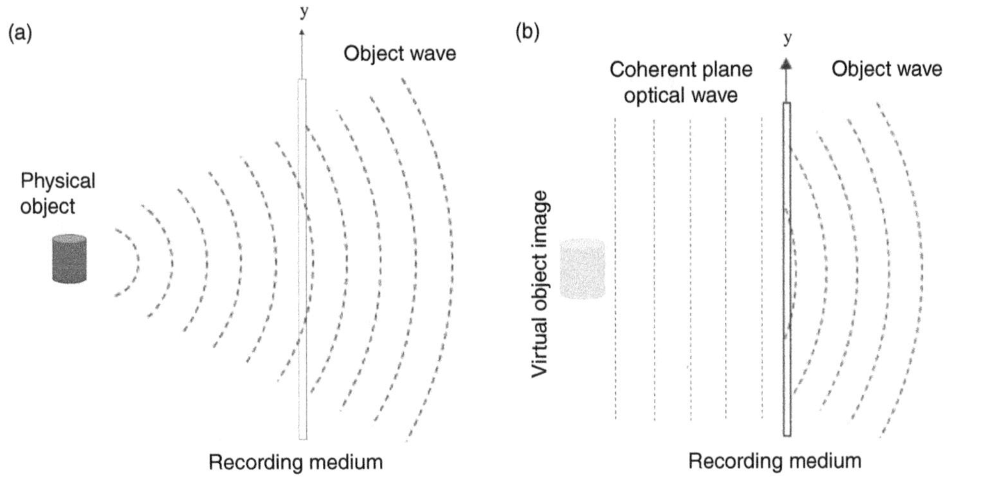

Figure 1.2 (a) Capturing object waves of an object on a recording medium. (b) Reconstructing a virtual object image from the recording medium with a coherent plane optical wave.

The magical recording media is known as a hologram, a word that finds its origin from the Greek words ὅλος (whole) and γραφή (drawing). Literally, the word hologram carries the meaning of something that is capable of recording the complete information of a visible image. Theoretically, there is little difference between looking at the virtual image of a hologram and the actual object. Hence, from a hologram, an observer should be able to see not only the color and intensity of the object, but also all the depth cues and disparity information. Depth cues are the visual information in an image that instigates the perception of distances between the observer and different parts of the object. For example, the lens of our eye will change its focus when objects at different distances are observed. Another example of depth cues is when the viewpoint of an observer is panned (shifted along a certain path): an object that is closer will appear to translate more than one that is farther away, which again creates the sensation of difference in depth between the objects. Disparity, sometimes referred as "binocular disparity," is the deviation in the location of an object as observed by the left and the right eyes. Through the disparity, our brain is able to interpret the depth of different objects in the scene.

The art of capturing a hologram of an object is commonly referred to as optical holography. The method was invented in late 1940 by Dennis Gabor, who was awarded the Nobel Prize in Physics in 1971 for his contribution to the invention and development of the holographic method. Works of Gabor in optical holography were reported in [1], and later by Bragg in [2]. There have been numerous advancements in optical holography since, and a comprehensive description on the developments can be found in [3,4]. For completion, a brief outline of holography is given here.

The concept of holography can be illustrated from a more analytical point of view. Figure 1.3 shows the side view of an object scene containing a single object point. The intensity of the object point is white, represented with a value of unity. A recording medium is positioned at the hologram plane to record the light wave of the object scene. As the white object point is illuminated with a coherent plane wave (i.e.,

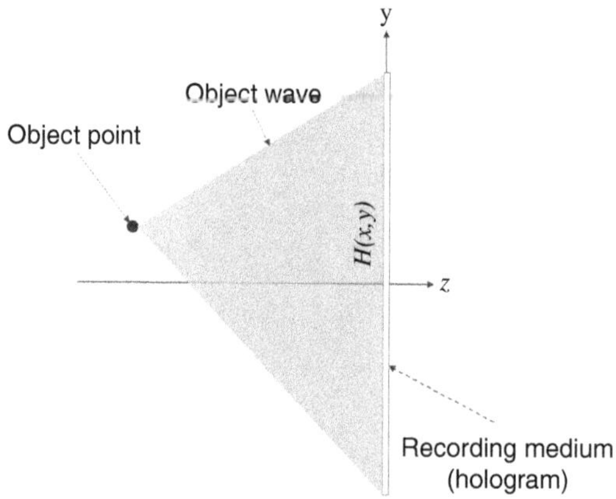

Figure 1.3 Object wave projected by an object point of unit intensity on the recording medium.

a wavefront that is homogeneous in magnitude and phase), it will behave as an isotropic point light source that scatters an object wave $O_0(x,y)$ onto the hologram. Extending this concept, for a single object point with intensity A, the object wave $O(x,y)$ that is projected on the hologram is given by

$$O(x,y) = A|O_0(x,y)|\exp[i\theta_0(x,y)]\exp[i\phi_0]. \tag{1.2}$$

In Eq. (1.2), the magnitude and phase of the object wave of a point source are denoted by $|O_0(x,y)|$ and $\theta_0(x,y)$, respectively. The term ϕ_0 is the phase of the light illuminating the object point, which will be added to the phase component of the object wave.

The above principles of hologram formation can be easily applied to a scene with multiple object points. An arbitrary object surface can be considered as comprising many isotropic point sources, each emitting its own object wave when illuminated. The optical signal recorded on the hologram is the superposition of the optical wave emitted from each object point. The mixing of these waves is a process that is commonly known as optical interference:

$$O(x,y) = \sum_{p=0}^{P-1} A_p|O_p(x,y)|\exp[i\theta_p(x,y)]\exp[i\phi_p], \tag{1.3}$$

where P is the total number of object points, and subscript p is the index of the object points. The object wave recorded on the hologram is referred as the interference pattern. For a coherent plane wave propagating along the axial direction (i.e., normal to the hologram plane), the phase ϕ_p of the illuminating wave is constant for all the object points, which can be ignored or included as a constant phase term in the computation of the object wave. If the illumination light beam is incoherent, the value of ϕ_p is random for different object points, and the object wave can no longer represent the light field of the source object.

For the time being, assume there are some means that enables the full object wave to be recorded on a medium, resulting in a hologram (denoted by the function $H(x,y)$ in

Figure 1.3). The signals on the hologram are complex-valued (with magnitude and phase components), and in the form of high-frequency fringe patterns. Different names have been adopted in various literatures to describe hologram signals, such as – but not limited to – fringe patterns, hologram fringes, diffractive waves, or holographic signals. These terms will be used throughout this book.

As shown in Figure 1.2(b), a coherent plane wave passing through a hologram is modulated, both in amplitude and phase to give the same object wave $O(x, y)$. From the reconstructed object wave, the observer will be able to see a realistic 3-D image of the object from the hologram. The process of recovering a 3-D image from the hologram is known as "reconstruction." The use of a coherent optical plane wave is mandatory as it will not change the amplitude and phase of the object waves – that is, A_p and $\theta_p(x, y)$ of each object point in both the recording and the reconstruction processes.

1.2 Optical Recording in Practice

A conceptual optical setup for recording the hologram of an object scene is shown in Figure 1.4. The actual recording system is slightly more complicated. A thin laser beam is converted into a plane wave with the use of a beam expander (BX), and illuminates the specimen after passing through the semi-transparent beam splitter (BS). A laser beam is used because it is coherent, so the phase distribution of its wavefront is homogeneous. As explained in Section 1.1, it is necessary that the phase ϕ_p of the illumination beam on every point on the object surface is constant. The beam expander magnifies the cross-sectional area of the thin laser beam so that the illumination light will be wide enough to cover the entire object of interest. The object wave emitted by the object of interest is partially reflected by the beam splitter to the recording media. If both the amplitude and phase of the object wave can be captured,

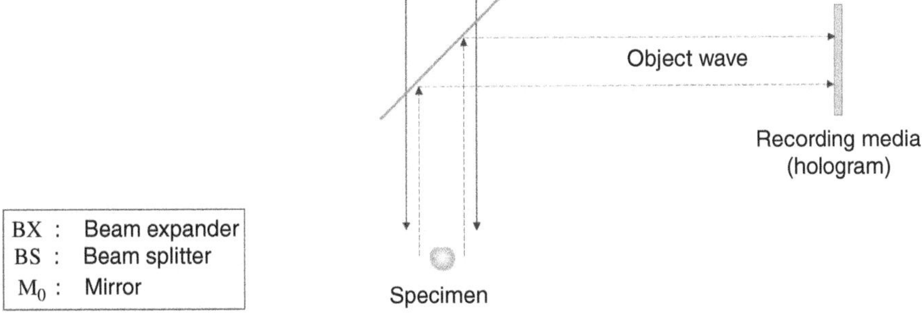

Figure 1.4 Conceptual optical setup in hologram recording.

the recording medium, will become the hologram of the specimen. Unfortunately, this ideal scenario is difficult, if not impossible, to achieve in practice.

The reason is that optical recording materials developed to date, commonly known as photographic films, are usually implemented with a layer of gelatin emulsion coated onto a plastic substrate. The emulsion comprises microscopic grains of silver halide crystal, each of which will be partially converted into silver upon exposure to light. Upon exposing the film to an optical image, the chemical property of each silver halide crystal will change proportionally to the intensity of light falling on it, forming an invisible or latent image. Subsequently, the film can be chemically developed into a visible image. Since the silver halide crystal is only responsive to the intensity of light, a photographic film is only capable of recording the intensity, but not the phase, of a light field. Instead of the full holographic information, the signal recorded on the photographic film is only constituted by the magnitude component of the object wave, as given by

$$O_I(x,y) = |O(x,y)|^2. \tag{1.4}$$

Apparently, the signal recorded by the photographic film in Eq. (1.4) is different from the object wave, as it does not carry any information on the phase component. Due to the absence of the phase terms, an image of the specimen cannot be reconstructed from the signal $O_I(x,y)$.

1.3 Photography

The classical art of photography can be interpreted as a special case of optical holography, although it was invented at a much earlier date when the art of holography was unknown to the world. As seen in Eq. (1.4), it is not possible to record the phase component of the object wave with a photographic film. Photography is based on capturing a small part of the object waves that carries the optical image of a specific viewpoint of the object scene. The concept is to place a small aperture between the object scene and the photographic film (a.k.a. a photograph), so that there is a one-to-one mapping between the light wave emitted from each object point and a single point on the recording plane. In another words, the light waves of different object points will not interfere with one another as they fall onto the recording media. Such an implementation is known as a "pinhole camera." Figure 1.5(a) shows the imaging of a pair of point light sources, A and B, with a pinhole camera. The tiny pinhole only allows a light wave of a unique orientation for each object point to pass through, and impinges on an image plane at which a photographic film is placed. Restricted by the pinhole, each location on a photograph can only receive the object wave of a single object point. Mathematically, the magnitude of the object wave received by a photograph from a single object point at location (x_p, y_p) is expressed as

$$O_P(x,y) = \begin{cases} |A_p|^2 & (x,y) = T(x_p,y_p) \\ 0 & \text{otherwise} \end{cases}, \tag{1.5}$$

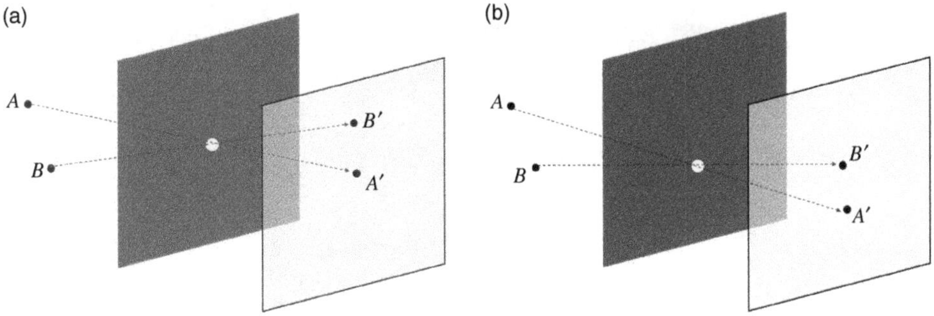

Figure 1.5 (a) A pinhole camera imaging a pair of points A and B. (b) Imaging points A and B with a pinhole at a different location.

where $T(x_p, y_p)$ denotes a one-to-one transformation on the coordinates (x_p, y_p). From Eq. (1.5), the function $O_P(x, y)$ is simply a transformed image of the object scene, with the object point recorded at the position $T(x_p, y_p)$. On the image plane, the image of an object point is flipped horizontally and vertically, and relocated to a new position according to the location of the pinhole. Note that the object wave contains only the magnitude component (i.e., the intensity) of the light wave of a unique point source on the photograph. As a result, the phase component of the light wave is not involved in the recording process, and the object can be illuminated with incoherent light (e.g., ambient lighting from the sun and fluorescent lamps), whereby the phase of the illumination beam on each object point can be totally random and uncorrelated with each other.

The photograph can only capture an image of the object scene at a particular viewpoint that is constrained by the position of the pinhole. With a single view of the object scene, the recorded image cannot convey any depth cue or disparity information. To capture another viewpoint of the object scene, the position of the pinhole has to be changed. Figure 1.5(b) shows the imaging of the pair of object points, A and B, with the pinhole shifted to a different location. From these two figures, the images of the two object points are relocated to different positions on the recording plane.

1.4 Recording Setup in Optical Holography

Although holographic technology can be traced back to 1940, hologram recording was not practiced until the development of the laser in the 1960s, which is capable of generating a coherent light wave. As mentioned before, coherent illumination is mandatory in holographic recording. The first hologram was demonstrated in 1962 by the Soviet scientist Yuri Denisyuk [5], as well as by Emmett Leith and Juris Upatnieks of the University of Michigan in the USA [6]. Different to photography, optical holography records a 3-D – instead of 2-D – image of the object scene onto a photographic film by preserving both the amplitude and phase components of the

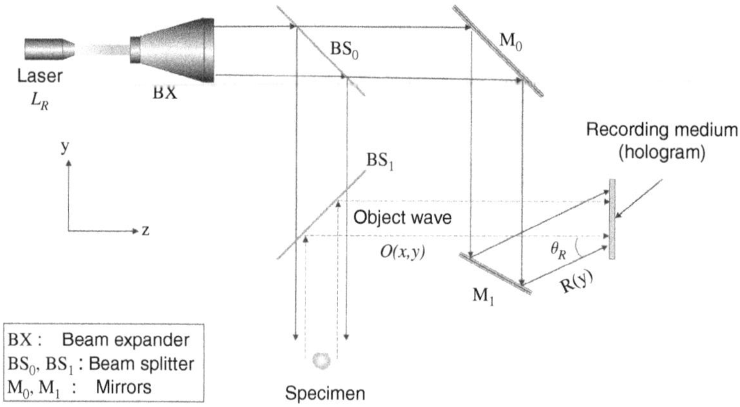

Figure 1.6 Practical optical setup for recording an off-axis hologram.

object wave. Intuitively, this may sound a bit contradictory, as a photographic film is only responsive to the intensity of the light wave. In holographic recording, this problem is overcome by converting the complex-valued holographic signal into intensity information prior to recording on a photographic film. The conversion is achieved by adding the complex-valued object wave with an inclined reference plane wave, which is also derived from the coherent beam that illuminates the object. The result of adding the pair of waves is an interference pattern (which is also commonly referred to as a beat pattern), encapsulating both the magnitude and phase components of the object waves. A typical optical setup for recording an off-axis hologram is given in Figure 1.6.

Similar to Figure 1.4, the source of illumination is a monochromatic laser beam L_R which is expanded into a plane wave via a beam expander. The expanded beam is divided into two separate paths after passing through a beam splitter. The first beam is reflected by the beam splitter BS_0, propagates through the beam splitter BS_1, and illuminates the specimen. The object wave $O(x, y)$ from the specimen is reflected by BS_1, and impinges on the hologram. The second part of the expanded beam is reflected to the hologram via mirrors M_0 and M_1, projecting the beam onto the hologram at an angle of incidence θ_R along the vertical direction, resulting in an inclined reference plane wave

$$R(y) = \exp i2\pi y\lambda^{-1} \sin(\theta_R), \tag{1.6}$$

where λ is the wavelength of the laser beam. Note that $R(y)$ is a pure phase function. If $\theta_R = 0$, the reference wave is simply the plane wave that is used to illuminate the specimen. When the object wave and the reference wave meet on the surface of the photographic film, they combine to form an interference pattern. The magnitude of the interference pattern is then recorded on the photographic film, resulting in a hologram with the intensity given by

$$
\begin{aligned}
H(x,y) &= |O(x,y) + R(y)|^2 \\
&= [|O(x,y)|^2 + |R(y)|^2] + O(x,y)R^*(y) + O^*(x,y)R(y) \\
&= C_1 + C_2 + C_3,
\end{aligned}
\tag{1.7}
$$

where $C_1 = [|O(x,y)|^2 + |R(y)|^2]$, $C_2 = O(x,y)R^*(y)$, and $C_3 = O^*(x,y)R(y)$.

From Eq. (1.7), the hologram $H(x,y)$ is a real signal that can be recorded onto the photographic film. Subsequently, the object wave can be optically reconstructed from the hologram by illuminating it with a plane wave, as shown in Figure 1.7. The light wave emitted by the hologram is contributed by three different components, each corresponding to a term in Eq. (1.7).

The first term C_1 is the sum of the intensities of the object and the reference beams. Due to the absence of the phase component, this part of the light wave does not carry useful information on the 3-D image of the specimen, and appears as a patch of highlighted region. The second term C_2 is identical to the object wave apart from the constant phase term $R^*(y)$, while the last term C_3 is the conjugate of the object wave multiplied with the reference wave $R(y)$. Being different from C_1, the complex object wave of the specimen are encapsulated in both C_2 and C_3.

From the optical wave of the illuminated hologram, three images can be observed. The first one is a focused, virtual image of the object scene that is reconstructed from the component C_2. This component of the optical wave contains the object wave $O(x,y)$ in the hologram $H(x,y)$, with the addition of the phase term $R^*(y)$. The virtual image can be viewed from an oblique angle, as shown in Figure 1.7. The second image is a blurred, defocused twin image corresponding to C_3, which carries $O^*(x,y)$, the conjugate of the object wave, and the phase term $R(y)$. Similar to the virtual image, the twin image can be observed from another oblique angle. The third image, usually referred to as the zero-order beam, is constituted from the remaining term C_1, and can be observed from the axial direction. Apart from the virtual image, the other pair of images are unwanted content as they do not form a correct reconstruction of the object scene.

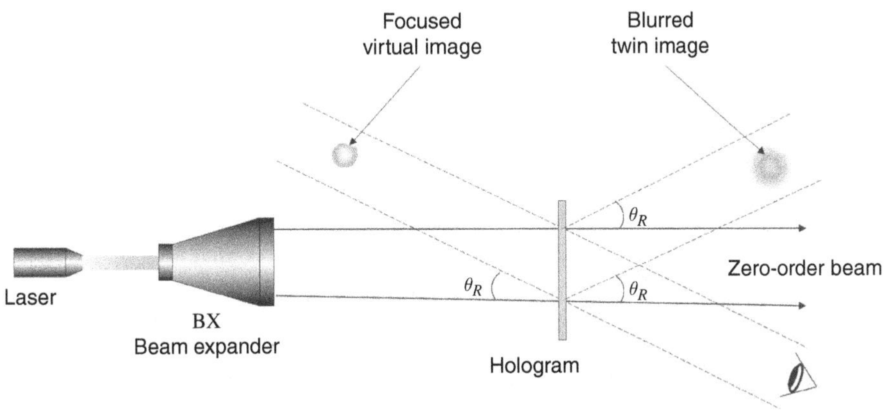

Figure 1.7 Optical reconstruction of an off-axis hologram.

From Figure 1.7, it can be seen that mixing of the object wave $O(x,y)$ with the inclined reference wave $R(y)$ has imposed an angular separation θ_R between each consecutive pair of images that are reconstructed from the hologram. It can be inferred that if θ_R is small, the virtual image could be partially overlapping with, and hence contaminated by, the twin image and the zero-order beam. As such, the angular separation θ_R must be large enough that the virtual image can be separated from the unwanted content, and optically reconstructed as a standalone visible image. Even if this is achieved, there is still a good chance that all three components are visible to the observer, although they are not spatially overlapping with one another. This is one of the major disadvantages of an off-axis hologram.

Theoretically, the virtual image can be fully separated from the unwanted content by increasing the value of θ_R. In practice, due to the limited resolution of the photographic film, a ceiling on the angle θ_R is imposed. As explained previously, a photographic film comprises individual microscopic grains of photosensitive materials, each of which constitute a pixel on the film. The resolution of the photographic film, therefore, is limited by the size of the pixel. Assuming that all the pixels are square in shape and have identical size of $\delta_d \times \delta_d$, an image recorded on it is effectively discretized with a 2-D sampling lattice with a sampling interval δ_d along the horizontal and vertical directions. According to sampling theory, this will impose an upper limit to the spatial frequency of the recorded signal.

To determine the maximum angle of incidence, consider a simple case where only the reference wave is present. For a given θ_R, the spatial frequency of the hologram fringe pattern is the minimum, as compared to the case when the object wave is present. Figure 1.8 shows the side view of a reference wave which is inclined at an angle θ_R along the vertical axis. As the signal on the hologram is discretized into pixels when it is recorded on the photographic film, the vertical axis is denoted by the variable n, where n is an integer representing the row indices of the pixels. The physical location of the kth row along the vertical axis is given by $y = k\delta_d$. A pair of light beams of the reference wave is shown in Figure 1.8, impinging on the hologram at $n = 0$ and $n = k$. The relative phase difference between these two beams is given by

$$d_p = 2\pi\lambda^{-1}d = 2\pi\lambda^{-1}k\delta_d \sin(\theta_R) = \omega k, \tag{1.8}$$

where $\omega = 2\pi\delta_d\lambda^{-1} \sin(\theta_R)$ rad/s. From Eq. (1.8) it can be visualized that the fringe pattern on the hologram is a sinusoidal wave with frequency ω. As the maximum frequency that can be represented in the discrete signal is π rad/s,

$$2\pi\delta_d\lambda^{-1} \sin(\theta_R) \leq \pi \Rightarrow \quad \sin(\theta_R) \leq 0.5\lambda\delta_d^{-1}. \tag{1.9}$$

For small value of θ_R, this expression can be approximated as

$$\theta_R \leq 0.5\lambda\delta_d^{-1}. \tag{1.10}$$

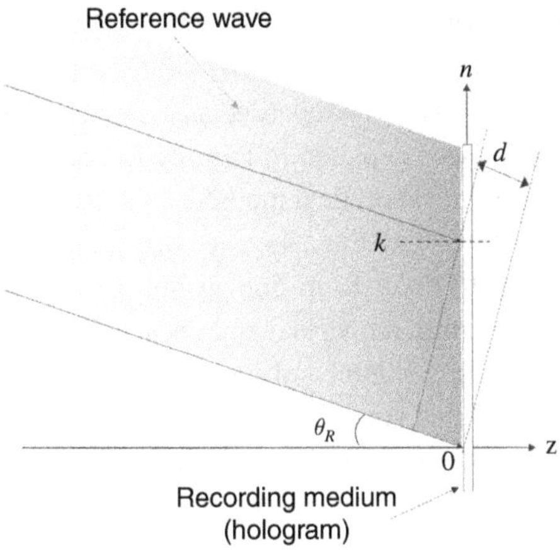

Figure 1.8 An inclined reference wave.

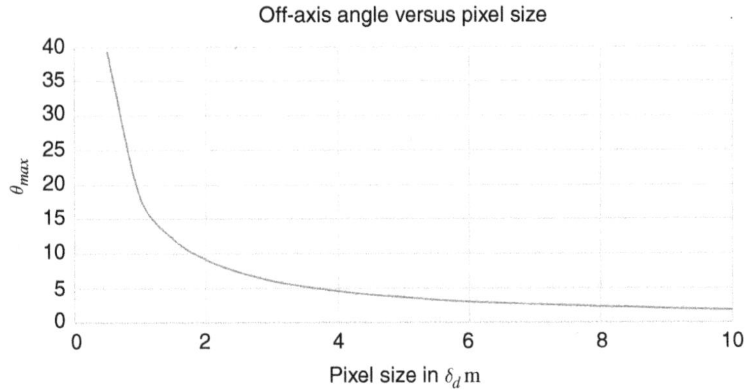

Figure 1.9 Plot of θ_{max} versus pixel size of the hologram.

In Eq. (1.10), a hologram with pixel size of δ_d can only record an incline plane wave with a maximum angle of incidence $\theta_{max} = 0.5\lambda\delta_d^{-1}$. This implies that the maximum viewing angle of this hologram is also limited to the angle θ_{max}.

A plot of the angle θ_{max} versus the pixel size δ_d of the hologram is shown in Figure 1.9. It can be seen that θ_{max} is inversely proportional to the pixel size of the hologram. Hence, the higher the resolution of the hologram, the larger will be the angular separation between the virtual image, the zero-order beam, and the twin image.

In optical holography, the hologram of a physical 3-D object is recorded in a photographic film. With the advancement of computing and electronics technologies, a digital hologram can be generated numerically from a computer graphic model, or captured from a physical object scene with a digital camera. In the following part of this chapter, the principles of operations of these two methods are described.

1.5 Computer-Generated Holography

Computer-generated holography (CGH) can be regarded as a computer simulation on the optical hologram-capturing processes depicted in Figures 1.4 and 1.6. Pioneering works on CGH can be traced back to the theoretical methods of J. Waters [7], and later production of the first binary computer-generated Fourier hologram by B.R. Brown and A.W. Lohmann [8–10]. Being different from optical holography, all entities in CGH, such as light, object, and hologram, are virtual and encapsulated as numbers or mathematical functions. For example, reference wave $R(y)$ is represented by the expression in Eq. (1.6), and its associated set of values λ and θ_R. As shown in Figure 1.10, the object scene is a 3-D computer graphic (CG) model of one or more physical objects. In general, the CG model can be defined by a set of object points that are located at certain positions in 3-D object space. It is often assumed that all the object points are illuminated with a uniform plane wave (which may be inclined at an angle θ_R), so that each object point will scatter a spherical light wave with intensity governed by its material properties. Such object points are sometimes referred to as "self-illuminating," and if an inline plane wave with $\theta_R = 0$ is assumed, the illuminating light beam can be discarded in the simulation.

The set of self-illuminating points in the CG model is input into a computer, in which a software program is applied to compute a digital hologram of the CG model, and the result is stored in a data file. At this point, the data file is merely a set of numbers that does not provide a lot of clues about the object scene. As such, the data file is converted into a physical hologram by either printing it on a static medium (e.g., a photographic film) or displaying it on an electronic accessible device such as a spatial light modulator (SLM). An SLM is a device that is capable of spatially modulating the optical properties of light falling on or reflecting from it. Display chips such as liquid

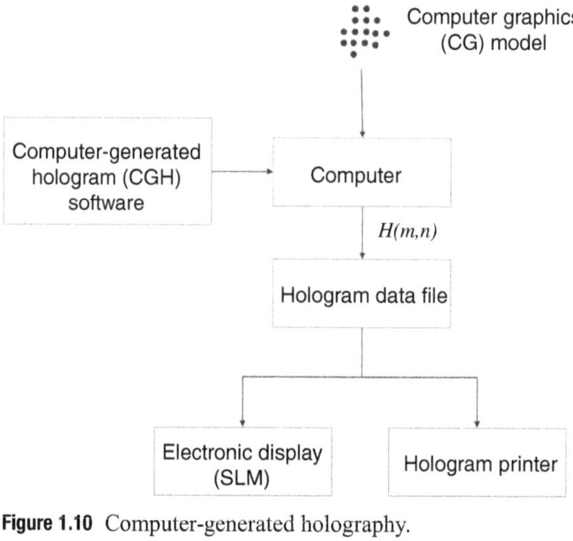

Figure 1.10 Computer-generated holography.

Table 1.1 List of essential parameters

Parameter	Symbol
Wavelength of light	λ
Sampling interval along the horizontal and vertical axes	δ_d
Horizontal extent of hologram and object space	H_DIM
Vertical extent of hologram and object space	H_DIM
Discrete horizontal coordinate	m
Discrete vertical coordinate	n
Axial coordinate (depth)	z

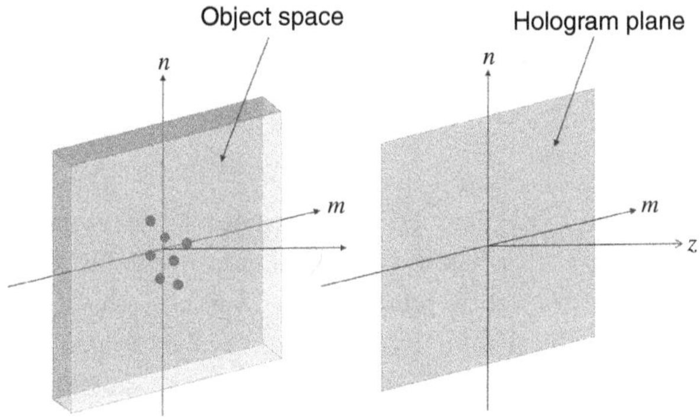

Figure 1.11 Spatial relation between the object space and the hologram plane.

crystal (LC) and liquid crystal on silicon (LCoS) are typical examples of SLMs. In any case, the 3-D object scene will be reconstructed as a visible image that can be observed by our eyes when the hologram is illuminated with a coherent beam. The process of converting a 3-D CG model into a digital hologram is referred to as computer-generated holography.

The fundamental principles of CGH are introduced in this section. Euclidean space, terminologies, and parameters, which will be referenced throughout the book, are defined. Euclidean space, as shown in Figure 1.11, defines a 3-D virtual environment that houses the object scene and the hologram. There are two major parts in the environment: the object space and the hologram plane. The object space contains the 3-D object scene (i.e., the CG model) which is composed of one or more objects, while the hologram plane contains the digital hologram to be generated. Major parameters that are frequently used throughout the book are defined in Table 1.1, and outlined as follows. Locations in Euclidean space are addressed with discrete rectangular Cartesian coordinates, with m and n (both of which are integers) denoting the discrete lateral (i.e., horizontal and vertical) coordinates. The axial direction, or the depth, that is orthogonal to the m and the n axes, is represented by the variable z. The hologram is assumed to be square in shape, with an identical number of rows and columns equal to H_DIM. The sampling intervals of the hologram and the lateral plane of the object space are identical and denoted by δ_d. Individual points in a hologram or an object scene are sometimes referred to as pixels. The physical location (x, y) of a discrete coordinate at (m, n) is related by $(x, y) = (m\delta_d, n\delta_d)$. The wavelength of light is denoted by λ.

Similar to optical holography, numerous research works have been conducted in CGH. There are two basic approaches in CGH, which are commonly known as the point-based method and the layer-based method. Both methods are straightforward and easy to implement, but the computation time is extremely lengthy. However, the latter part of this book will explain how this problem can be overcome with fast algorithms, leading to significant shortening of the computation time. As the point-based and layer-based methods are both important foundations of CGH, they will be described in more detail in the subsequent sections. In passing, there are other, more sophisticated techniques in CGH, such as polygon-based methods. However, they are beyond the scope of this book; interested readers may refer to [11–13] for details.

1.5.1 Point-Based Method

In the point-based method [14], the 3-D object scene is represented by a set of points $\Lambda = \{o_0, o_1, \ldots \; o_{P-1}\}$, where P is the total number of object points. In some literature, where the object scene is presented as an image, object points are also referred to as pixels. Each object point in Λ is indexed by the variable p and characterized by its intensity A_p, its position (m_p, n_p) on the lateral plane, and its axial distance z_p from the hologram plane. The axial distance is often referred to as the depth of the object point. For a single point with unity intensity, and located at m_p, n_p, and depth $= z_p$, its object wave impinging on the hologram can be obtained with Fresnel diffraction as given by

$$O_p(m, n, m_p, n_p;\; z_p) = r^{-1}_{m;n;m_p;n_p;z_p} \exp[(i2\pi\lambda^{-1} r_{m;n;m_p;n_p;z_p})], \tag{1.11}$$

where $r_{m;n;m_p;n_p;z_p} = \sqrt{(m - m_p,\;)^2\delta_d^2 + (n - n_p)^2\delta_d^2 + z_p^2}$ is the distance between an object point at (m_p, n_p) on the image plane at depth $= \; z_p$, to a pixel at (m, n) on the hologram.

If the object point is located at the origin (i.e., $m_p = n_p = 0$), the object wave is reduced to a more simple expression given by

$$\begin{aligned} fzp(m, n; z_p) &= O_p(m, n, 0,\; 0;\; z_p) \\ &= r^{-1}_{m;n;0;0;z_p} \exp[(i2\pi\lambda^{-1} r_{m;n;0;0;z_p})], \end{aligned} \tag{1.12}$$

a function that is commonly known as the Fresnel zone plate (FZP). If the separation between the object space and the hologram plane is sufficiently large (i.e., $z_p \gg M; N$), the term $r^{-1}_{m;n;0;0;z_p}$ can be approximated as a constant term and discarded in the expression. In any case, it can be easily inferred from Eq. (1.12) that the function $O_p(m, n, m_p, n_p; z_p)$ is simply a translated version of the FZP, that is

$$O_p(m,n,m_p,n_p;\ z_p) = fzp(m - m_p, n - n_p; z_p). \tag{1.13}$$

If the object scene is represented by the set of points Λ, the object waves on the hologram plane can be computed as the superposition of the object waves of all the object points, each weighted by its intensity value A_p as

$$\begin{aligned} H(m,n) &= \sum_{p=0}^{P-1}[A_p \times O_p(m,n,m_p,n_p;z_p)] \\ &= \sum_{p=0}^{P-1}[A_p \times fzp(m - m_p, n - n_p; z_p)]. \end{aligned} \tag{1.14}$$

The function $H(m,n)$ that records the collection of the object waves is known as a digital Fresnel hologram of the object scene, as it is obtained from the Fresnel diffraction equation in Eq. (1.11). Referring to the above formulation, the hologram is a complex-valued function. As the hologram is generated in a point-by-point manner, the process is generally known as the point-based method.

To illustrate the point-based method, a numerical simulation of generating a digital Fresnel hologram of a set of object points is demonstrated with the MATLAB codes in Section 1.8.1. In the simulation program the object space is derived from a source image comprising a pair of characters, "H" and "K," having brightness of 255 and 128, as shown in Figure 1.12(a). Each non-zero pixel in the image is taken to be an object point. The image is evenly partitioned into a left half-plane, and a right half-plane that are located at 0.07 m and 0.08 m from the hologram, as illustrated in Figure 1.12(b). In another words, the object space is a double-depth image. The size of the hologram is 256 × 256. The optical settings, which will be adopted in all the simulation programs in this book, are shown in Tables 1.2.

After running the main program, the Fresnel hologram is generated; the real and the imaginary parts of the hologram are displayed in Figure 1.13. It can be seen that the

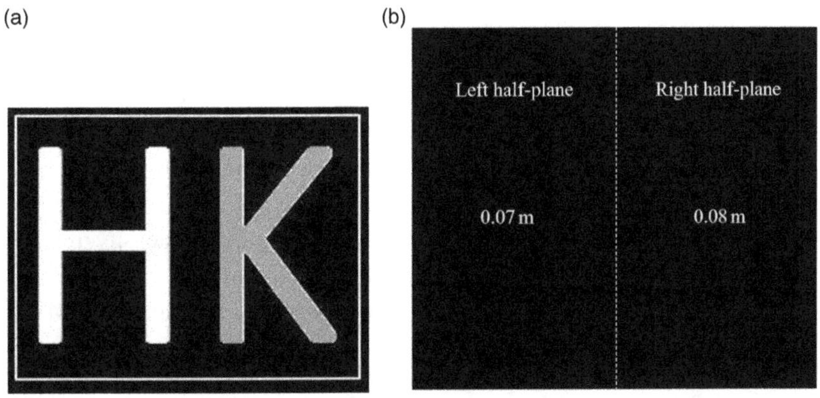

Figure 1.12 A double-depth image comprising the letters H and K. The image is evenly partitioned into a left half-plane and a right half-plane, each being a layer of the object space. (a) Intensity of the double-depth image; (b) distances of the two layers from the hologram.

Table 1.2 Optical settings adopted in the simulation programs of this book

Wavelength λ	520 nm
Pixel size of hologram δ_d	6.4 μm

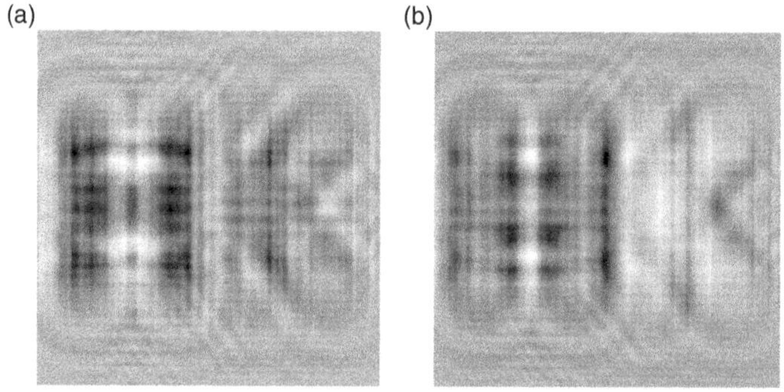

Figure 1.13 A digital Fresnel hologram of the double-depth image shown in Figure 1.12, generated with the point-based method and based on the optical settings in Table 1.2. (a) Real part of the hologram. (b) Imaginary part of the hologram.

holographic images in both components are composed of high-frequency fringes that bear little resemblance to the original object image.

1.5.2 Layer-Based Method

Although the point-based method is effective in generating the digital Fresnel hologram of a 3-D object scene, the computation will be extremely heavy if there are a lot of object points. For example, if the object space is a planar image of size 256 × 256, there are over 64 000 object points, and the object wave of each of them has to be calculated and superimposed on the hologram. This problem can be alleviated with the layer-based method.

In the layer-based method, the 3-D object space is evenly partitioned into a sequence of D depth planes, denoted by $\Gamma_S = I_j(m,n)|_{1\leq j\leq D}$, which are parallel to the hologram as shown in Figure 1.14. Each depth plane is a 2-D image representing the intensity and spatial distribution of the object points that are residing on that plane, which are at the same depth z_j from the hologram. If the intensity of the entire image is zero at a particular location on a depth plane, it means that no object point is there.

Applying the concept of the point-based method, the object wave of the object points on the jth depth plane is

$$O_j(m,n) = \sum_{u=0}^{M-1}\sum_{v=0}^{N-1}[I_j(u,v) \times fzp(m-u, n-v; z_j)]. \tag{1.15}$$

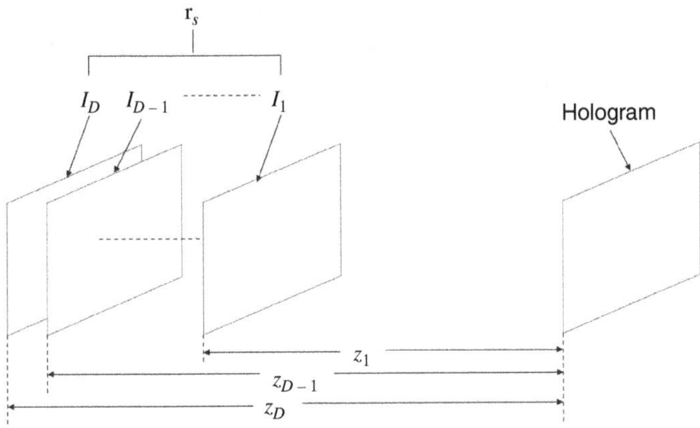

Figure 1.14 Partitioning the object space into a sequence of evenly spaced depth planes which are parallel to the hologram.

Equation (1.15) can be rewritten as a convolution process as

$$O_j(m,n) = I_j(m,n) * fzp(m,n;z_j). \tag{1.16}$$

The Fresnel hologram of the object scene is obtained from the sum of the object waves of all the depth planes, which can be expressed as

$$H(m,n) = \sum_{j=1}^{D} O_j(m,n). \tag{1.17}$$

Replacing the arithmetic calculations in Eq. (1.15) with the convolution operation in Eq. (1.16) does not lead to any reduction in the amount of computation. However, to be described later, convolution can be speeded up significantly if it is implemented in the frequency instead of the spatial domain. In this respect, convolution has a definite advantage over the point-based method in generating a hologram. Besides, convolution can also lead to some convenience in software implementation. Apart from convolution, there are other improved techniques for applying the layer-based method, such as, among others, [15–17]. From Eq. (1.17) it can be inferred that the computation time will become more and more significant with increasing number of depth planes, as well as the size of the object space and the hologram. As such, the computation advantage of the convolution method is reduced in generating holograms of thick objects, in which case the depth range has to be partitioned into a large number of layers.

1.6 Reconstruction of Digital Hologram

The process of recovering a visual image of the 3-D object space from a hologram is known as reconstruction [18]. When a physical hologram (that is printed on a recording medium or displayed on an electronic accessible device) is illuminated with a coherent beam, a visible image of the 3-D object space can be observed from the hologram. This is generally referred to as optical reconstruction. For a digital hologram, reconstruction can be realized with computation, a process that is generally known as numerical reconstruction.

Hologram reconstruction can be interpreted as the reverse process of hologram generation, where the light beams are propagated from the hologram plane to the object space. This can be quite apparent in the case of an object scene with a single-depth plane at $z = z_1$. Applying the layer-based method in Eq. (1.16), the Fresnel hologram $H(m,n)$ corresponding to this single plane object is the convolution between the intensity image, and the FZP is

$$H(m,n) = O_1(m,n) = I_1(m,n) * fzp(m,n;z_1). \tag{1.18}$$

By inspection, the intensity image can be recovered from the hologram by convolving it with the conjugate of the FZP:

$$\begin{aligned} &H(m,n) * fzp^{\dagger}(m,n;z_1) \\ &= [I_1(m,n) * fzp(m,n;z_1)] * fzp^{\dagger}(m,n;z_1) \\ &= I_1(m,n). \end{aligned} \tag{1.19}$$

The same principle can be extended to the reconstruction of a hologram representing an object scene with multiple depth planes. Refer again to Eqs. (1.16) and (1.17) on a hologram that is generated with the convolution-based method. To reconstruct the image $I_{R;k}(m,n)$ located at the focused plane $z = z_k$, Eq. (1.19) is applied to convolve the digital hologram with the complex conjugate of the FZP at the corresponding depth. Mathematically this is presented as

$$\begin{aligned} I_{R;k}(m,n) &= \sum_{j=1}^{D} O_j(m,n) * fzp^{\dagger}(m,n;z_k) \\ &= \sum_{j=1}^{D} [I_j(m,n) * fzp(m,n;z_j) * fzp^{\dagger}(m,n;z_k)] \\ &= I_k(m,n) + \sum_{\substack{j=1 \\ j \neq k}}^{D} [I_j(m,n) * fzp(m,n;z_j) * fzp^{\dagger}(m,n;z_k)] \\ &= I_k(m,n) + J_R(m,n). \end{aligned} \tag{1.20}$$

It can be seen that the image $I_k(m,n)$ at the focused depth plane $z = z_k$ is fully recovered in the reconstructed image, while the rest of the depth planes are reconstructed as blurred, defocused content $J_R(m,n)$. This is similar to how a 3-D scene is observed with eyes. When focusing on a near subject, those that are farther away will appear to be blurred, and vice versa.

As an example, a digital Fresnel hologram of the double-depth image in Figure 1.12 is generated with the layer-based method. From the generated hologram, the object scene, which is composed of two layers, is reconstructed from it in a plane-by-plane manner. The reconstructed images at the two depth planes are shown in Figures 1.15. In both cases, the character that is located at the focused plane is reconstructed as a sharp image, while the other is a blurred image. Apart from slight blurriness, the character that is reconstructed at its focused plane is similar to the original image.

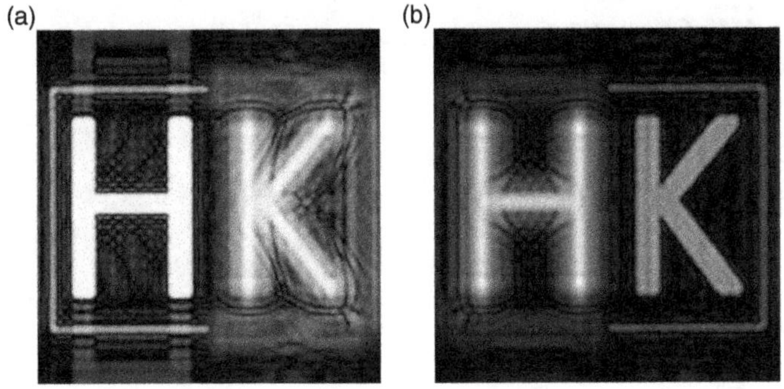

Figure 1.15 (a) Reconstructed image at depth plane 0.07 m. (b) Reconstructed image at depth plane 0.08 m.

These results reflect the capability of a hologram in preserving the intensity and depth information of a 3-D object scene.

1.7 Capturing Digital Hologram of a Physical Object

Although CGH can be employed to generate a hologram of a CG model, it cannot be applied directly – as in the case of optical holography – in capturing the digital hologram of a physical object. Referring to Figure 1.4, a digital hologram can be acquired by simply replacing the photographic film with a digital CCD camera. Similar to a photographic film, a CCD camera can only record intensity information. Hence, the off-axis hologram setup shown in Figure 1.6 is adopted in the capturing of a digital hologram. Since a digital camera comprises of a 2-D array of discrete optical sensors, the image recorded is also discretized in the same way. As such, the digital hologram $H(x, y)$ will be replaced by its discrete form $H(m, n)$, where m and n are integers representing the discrete coordinates along the horizontal and vertical directions. The following subsections introduce four methods for capturing digital holograms of a physical 3-D object.

1.7.1 Capture of Digital Off-Axis Hologram

Section 1.4 and Figure 1.7 infer that a digital off-axis hologram will suffer the same disadvantage as an optical off-axis hologram, with the virtual image being disrupted by the zero-order beam and the twin image. If the resolution of the digital hologram is not high enough, these three components will overlap each other, and a virtual image of the object space cannot be reconstructed correctly. Being different from a photographic film, which can have a very high resolution of less than 1 μm, the pixel size of a CCD camera is relatively much coarser. For example, the pixel size of a typical high-end scientific camera is around 5–7 μm. From the chart in Figure 1.9, the maximum angle of diffraction θ_{max} is limited to the range $[2.6^{\circ}, 3.6^{\circ}]$. Nevertheless, the off-axis arrangement is by far the most convenient way of acquiring a digital hologram.

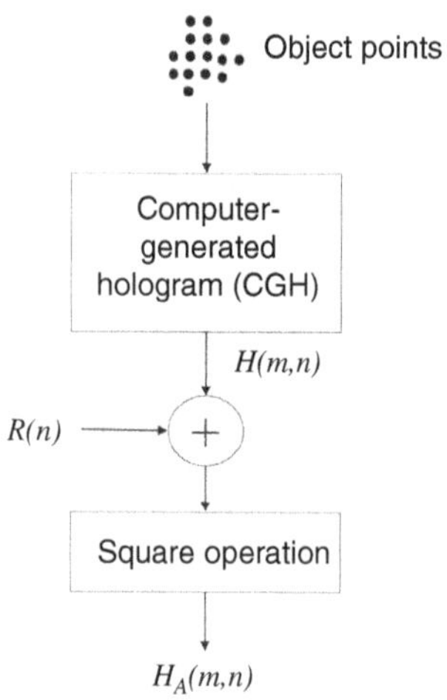

Figure 1.16 Generation of digital off-axis hologram.

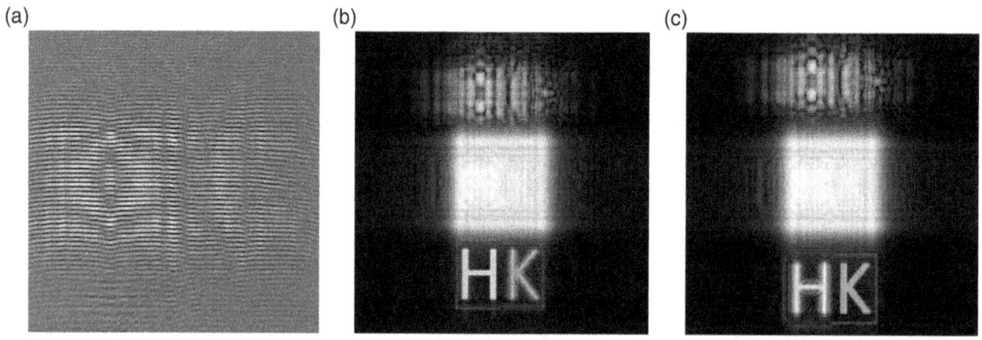

Figure 1.17 (a) Off-axis hologram of a double-depth image with $\theta_R = 1.2°$. (b, c) Reconstructed image of the off-axis hologram at depth planes 0.07 m and 0.08 m, respectively.

Mathematically, the mechanism of capturing an off-axis hologram of a 3-D object scene is given by Eq. (1.7), and shown in the block diagram in Figure 1.16. A simulation of the process is provided in the MATLAB program given in Section 1.8.2.

In the simulation, a digital Fresnel hologram of the double-depth image in Figure 1.12 is first generated with the layer-based method. After the hologram has been generated, a reference plane wave which is inclined at an angle of $\theta_R = 1.2°$ along the vertical direction is added to the hologram. The square of the magnitude of the mixed waves is taken to be the off-axis hologram, as shown in Figure 1.17(a). Next, the images at the two focused planes are reconstructed from the off-axis hologram, as shown in Figures 1.17(b) and 1.17(c). In each figure, it can be seen that the image of the character which is located at the focused plane is reconstructed clearly, while the other character becomes a defocused image. The bright square patch at the center is the zero-order beam, and the blurred region above it is twin image.

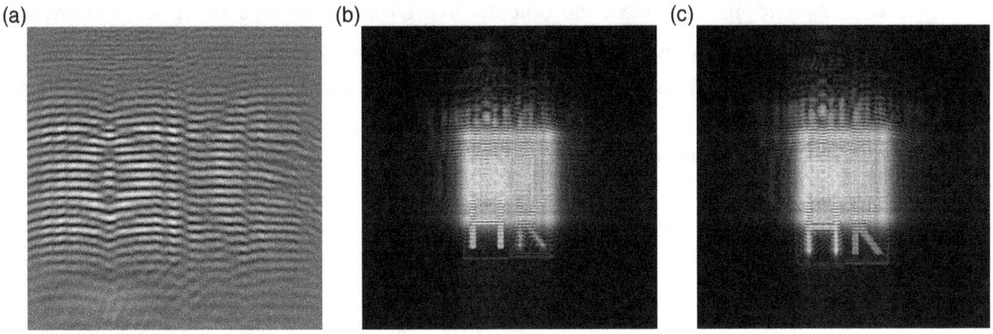

Figure 1.18 (a) Off-axis hologram of a double-depth image with $\theta_R = 0.6°$. (b, c) Reconstructed image of the off-axis hologram at depth planes 0.07 m and 0.08 m, respectively.

A similar simulation is performed with the incidence angle θ_R being changed to 0.6°. The off-axis hologram is shown in Figure 1.18(a), and the reconstructed images at the two focused planes are shown in Figures 1.18(b) and 1.18(c). It can be seen that when the angle θ_R is small, the virtual image, twin image, and zero-order beam are overlapping and cannot be discriminated clearly.

1.7.2 Phase-Shifting Holography

It is desirable to capture both the amplitude and phase of the interference pattern of a hologram with a digital camera, so that the object scene can be fully reconstructed without being contaminated with the zero-order beam and the twin image. As mentioned previously, a digital camera can only record intensity information. This problem has been overcome with the phase-shifting holography (PSH) technique, which was first introduced in [19] for capturing electronic holographic signals with camera tubes.

In the basic PSH hologram acquisition method, an object specimen is illuminated by a coherent plane wave, which can be generated by expanding a laser beam with a beam expander. The object wave of the illuminated specimen (which is a complex-valued function) is sequentially added with four different phase-shifted versions of a reference wave that are derived from the illumination beam. The summation of the object wave with each version of the reference wave results in a beat, or interference pattern. The magnitude of each interference pattern, referred to as an interferogram, is captured by a camera. Subsequently, the complex-valued object wave of the specimen can be recovered from the four interferograms. Since four exposures are required to capture the interferograms, the method is known as four-step PSH.

The general optical setup of a PSH system is illustrated in Figure 1.19. A laser beam is expanded by the beam expander into a plane wave $\exp(i\phi_R)$ and split into two light beams by the beam splitter BS_1. The first beam is reflected by the mirror M_1, and directed to illuminate the specimen. The second beam propagates through a phase shifter (PS), whereby one of the four phase-shift values in the set $PH_4 = \{0, \pi/2, \pi, 3\pi/2\}$ is added to form a phase-shifted reference plane wave

$\exp(i\phi_R + i\phi_k)|_{\phi_k \in PH}$. Next, the object wave $O(m,n)$ emitted from the specimen is sequentially mixed with each of the reference beams, and the magnitude of the interference pattern (i.e., the interferogram) is recorded by the digital camera. Let $|O(m,n)|$ and $\theta(m,n)$ denote the magnitude and phase of the object wave, respectively; the four interferograms are

$$\begin{aligned} IN_1(m,n) &= |O(m,n) + \exp(i\phi_R)|^2 \\ &= C + 2|O(m,n)| \cos\left(\theta(m,n) + \phi_R\right) \end{aligned} \tag{1.21}$$

$$\begin{aligned} IN_2(m,n) &= |O(m,n) + \exp(i\phi_R + i\pi/2)|^2 \\ &= C + 2|O(m,n)| \cos\left(\theta(m,n) + \phi_R + \pi/2\right) \end{aligned} \tag{1.22}$$

$$\begin{aligned} IN_3(m,n) &= |O(m,n) + \exp(i\phi_R + i\pi)|^2 \\ &= C + 2|O(m,n)| \cos\left(\theta(m,n) + \phi_R + \pi\right) \end{aligned} \tag{1.23}$$

$$\begin{aligned} IN_4(m,n) &= |O(m,n) + \exp(i\phi_R + i3\pi/2)|^2 \\ &= C + 2|O(m,n)| \cos\left(\theta(m,n) + \phi_R + 3\pi/2\right), \end{aligned} \tag{1.24}$$

where $C = \left(1 + |O(m,n)|^2\right)$.

It can be easily proved that

$$\begin{aligned} &\frac{1}{4}\left(IN_1(m,n) - IN_3(m,n)\right) + \frac{i}{4}\left(IN_2(m,n) - IN_4(m,n)\right) \\ &= |O(m,n)|\exp\left(i\theta(m,n)\right)\exp(i\phi_R) = O(m,n)\exp(i\phi_R). \end{aligned} \tag{1.25}$$

Apart from the additional constant phase term $\exp(i\phi_R)$, which can be neglected in practice, the original complex object wave of the specimen is recovered.

A simulation of the four-step PSH hologram acquisition process in Figure 1.19 is conducted with the code listed in Section 1.8.3. The source image is the double-depth

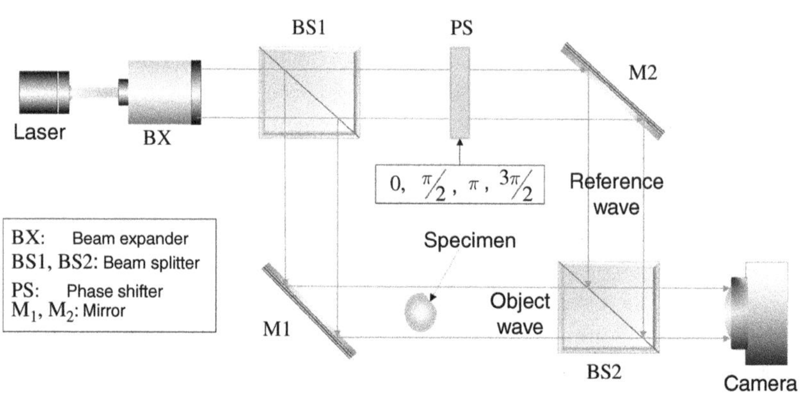

Figure 1.19 Optical setup of a four-step phase-shifting holography (PSH) system.

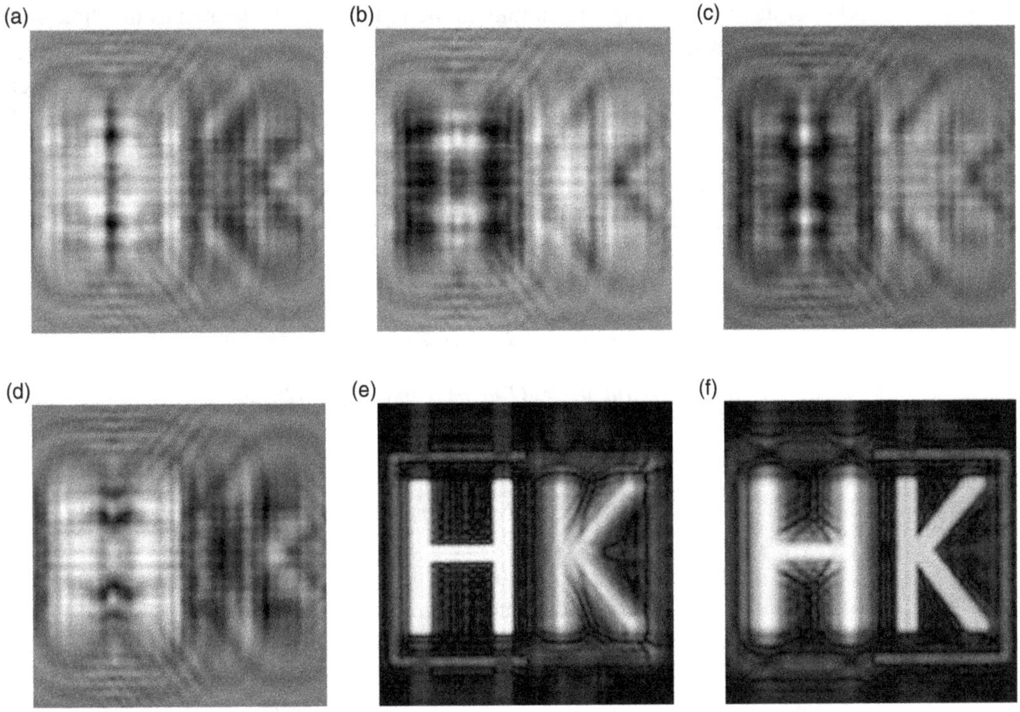

Figure 1.20 (a–d) The four interferograms corresponding to the hologram of the double-depth image in Figure 1.12. (e) Reconstructed image at depth plane 0.07 m. (f) Reconstructed image at depth plane 0.08 m.

image shown in Figure 1.12, containing the characters H and K located at distances of $z_1 = 0.07$ m and $z_2 = 0.08$ m from the hologram. First, a hologram representing the double-depth image is generated with the layer-based method. Next, four interferograms are generated, each by adding the hologram with one of the four phase-shifted versions of the reference wave, and taking the magnitude of the result. For simplicity, the phase of the reference wave before passing through the phase shifter (PS) is assumed to be zero. After executing the program, the four intensity images of the interferograms captured by the CCD camera are shown in Figure 1.20(a–d). Next, the hologram is composed from the four interferograms, and the reconstructed images at the two focused planes are derived from the hologram and shown in Figures 1.20(e) and 1.20(f).

Despite the effectiveness of the four-step PSH in capturing a complex-valued hologram of a physical 3-D object, the process is rather tedious as the four interferograms have to be captured. As each interferogram is formed by adding to the object wave a phase-shifted version of the reference wave, the capturing of the interferograms has to be conducted in a sequential manner. Over the years, quite a number of research investigations have been conducted to reduce the number of interferograms in a PSH system.

One of the early attempts to simplified the four-step PSH, known as the three-step PSH, was proposed by Y. Awatsuji et al. in [20]. In this approach, only three interferograms are captured, and these are integrated to recover the hologram of the 3-D object. The mechanism of the three-step PSH is similar to the four-step PSH, in which each interferogram is acquired by adding the object wave with one of the three phase-shifted versions, $PH_3 = \{0, 2\pi/3, -2\pi/3\}$, of the reference wave.

Mathematically, the three interferograms can be expressed as

$$IN_{3;1}(m,n) = |O(m,n) + \exp(i\phi_R)|^2 \tag{1.26}$$

$$IN_{3;2}(m,n) = |O(m,n) + \exp(i\phi_R + i2\pi/3)|^2 \tag{1.27}$$

$$IN_{3;3}(m,n) = |O(m,n) + \exp(i\phi_R - i2\pi/3)|^2. \tag{1.28}$$

It can be easily verified that

$$O(m,n) = IN_{3;1} + \exp(i2\pi/3)IN_{3;2}(m,n) + \exp(-i2\pi/3)IN_{3;3}(m,n). \tag{1.29}$$

Subsequent to the three-step PSH, a two-step PSH [21–22] has been developed that only requires the acquisition of two interferograms, and the intensities of the object and the reference waves. Later, J.-P. Liu and T.-C. Poon [23] proposed a simplified method that does not require the recording of the reference and the object wave intensities.

Despite the improvements in PSH technology over the years, it requires the sequential acquisition of at least two interferograms, hence limiting its application to the imaging of stationary or very slow-moving specimens. In the case of recording a hologram of a fast-moving object, it is mandatory that all the interferograms are acquired concurrently. In [24], a parallel phase-shifting hologram (PPSH) method, as shown in Figure 1.21(a), is proposed. The reference wave is impinged on a phase-shifting array that contains downsampled version of different phase-shifting patterns. A small section of a typical phase-shifting array corresponding to a four-step PSH is shown in Figure 1.21(b). When the reference beam illuminates a particular cell in the phase-shifting array, it will be phase-shifted by one of the four values $\left[0, \frac{\pi}{2}, \pi, \frac{3\pi}{2}\right]$. The optical wave that emerges from the phase-shifting array comprises all the four phase-shifted version of the reference waves, which are spatially multiplexed in a non-overlapping manner. On the sensing plate of the camera, the object wave of the specimen mixes with a different phase-shifted version of the reference waves, resulting in four downsampled interferograms.

Subsequently each interferogram, after applying interpolation to fill in the missing pixels, is combined with Eq. (1.25) to obtain the hologram of the specimen. As all the interferograms are captured in a single round, the PPSH method is sometimes referred to as a parallel or one-step PSH.

By the same principle, the PPSH method can also be adopted to implement the three-step [20] and two-step PSH methods [21]. For each of these configurations, a phase-shifting array comprising triplets or pairs of phase-shifting pixels is overlaid

(a)

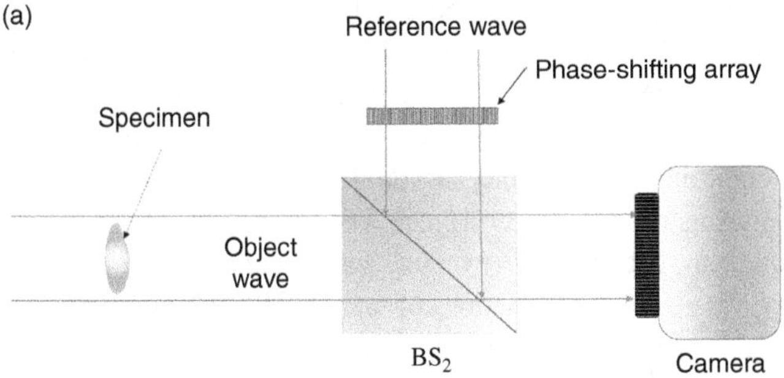

(b)

0	π / 2	0	π / 2
π	3π / 2	π	3π / 2
0	π / 2	0	π / 2
π	3π / 2	π	3π / 2

Figure 1.21 (a) One-step phase-shifting holographic system implemented with a phase-shifting array. (b) A small section of a four-step phase-shifting array.

onto the digital camera. With the PPSH, the interferograms are downsampled and need to be interpolated to their original size before being applied to compose the hologram. However, there are bound to be certain losses when the interferograms are downsampled, which cannot be fully recovered with interpolation. This will lead to distortion of the hologram composed from the interferograms, and hence its reconstructed image. The problem can be alleviated with a selective interpolation algorithm to include the sub-pixel data [25]. In comparison with linear interpolation, the selective interpolation method can lead to around 15% improvement of the reconstructed image of the hologram. Another interpolation method was proposed by Xia et al. [26], which is targeted at recovering the high-frequency signals in the downsampled interferograms. Later, the PPSH was applied in measuring the spectral reflectance of 3-D objects [27] and development of portable hologram-capturing systems [28,29].

The variants of the PSH methods have their pros and cons. Depending on the nature of the specimens and the precision required in capturing the holograms, different types of PSH methods may have to be used. For example, the single-step PSH is preferred for capturing the hologram of a dynamic object at the expense of introducing aliasing errors, while multiple-step PSH should be adopted for acquiring holograms with fewer errors. The above descriptions infer that switching between different PSH configurations will require changes in certain parts of the optical settings (e.g., replacing the phase-shifting array in changing a four-step PSH into a three-step PSH configuration). In view of this, a phase-mode SLM for generating the phase retardation patterns is proposed in [30].

1.7.3 Optical Scanning Holography

Sections 1.7.1 and 1.7.2 introduce two popular methods for recording holograms of physical objects. Both techniques are based on capturing, with a digital camera, the diffracted waves of the specimen that are illuminated with a coherent plane wave. A major disadvantage of these two methods is that the size and resolution of the hologram is limited by the receiving area of the image sensor. The coverage of the scene to be recorded, as well as the pixel size of the hologram, are therefore restricted by the image sensor.

The optical scanning holography (OSH) technique [31,32] developed by Poon and Korpel in the late 1970s, has overcome these restrictions. Based on a scanning mechanism and a single pixel sensor, the OSH is capable of capturing holograms of both microscopic and macroscopic objects. The full setup of an OSH system is rather complicated, and only a simplified concept is presented here. The hologram-capturing process of OSH can be divided into two stages. The first stage is shown in Figure 1.22(a), and explained as follows. The illumination beam $\Psi(x, y; z)$ is a combination of a coherent plane wave and a coherent spherical wave that is generated from a point source, both having an optical frequency ω_0. An acousto-optic modulator is used to modulate the frequency of the plane wave to $(\omega_0 + \Omega)$. For simplicity of description, the specimen is assumed to be a planar object that is parallel to the plane wave. Denoting the plane and the spherical waves by $A\exp[i(\omega_0 + \Omega)t]$, and $B\exp\left[\frac{-ik_0}{2z}(x^2 + y^2)\right]\exp[i\omega_0 t]$, respectively, we have

$$\Psi(x, y; z) = A\exp[i(\omega_0 + \Omega)t] + B\exp\left[\frac{-ik_0}{2z}(x^2 + y^2)\right]\exp[i\omega_0 t], \tag{1.30}$$

where Ω is known as the heterodyne frequency and z is the axial distance between the specimen and the point source that generates the spherical wave.

Utilizing an xy scanning device, the illumination beam is used to scan a specimen S along a zig-zag path. The optical waves scattered by the specimen are collected by a photo-detector through a focusing lens, resulting in a time-varying current $i(x, y)$. The zig-zag scanning of the specimen by the illumination beam can be described with a convolution process, as given by

$$i(x, y) = |T(x, y; z) * \Psi(x, y; z)|^2. \tag{1.31}$$

After band-pass filtering with a filter tuned at the heterodyne frequency Ω, a signal $i_\Omega(x, y)$, which is known as the heterodyne current, results as

$$i_\Omega(x, y) = AB \sin\left[\Omega t + \frac{k_0}{2z}(x^2 + y^2)\right] * |T(x, y; z)|^2. \tag{1.32}$$

The second stage of an OSH system, which is used to extract the real and the imaginary components of the hologram of the specimen from the heterodyne current, is shown in Figure 1.22(b). The heterodyne signal $i_\Omega(x, y)$ is input to a pair of mixers that multiply it with a cosine and a sine function. After mixing, the signals are low-pass filtered to give a signal $i_c(x, y)$ and a signal $i_s(x, y)$, as given by

$$i_c(x, y) = LPF[i_\Omega(x, y) \times \sin(\Omega t)] = K \cos\left[\frac{k_0}{2z}(x^2 + y^2)\right] * |T(x, y; z)|^2,$$

and

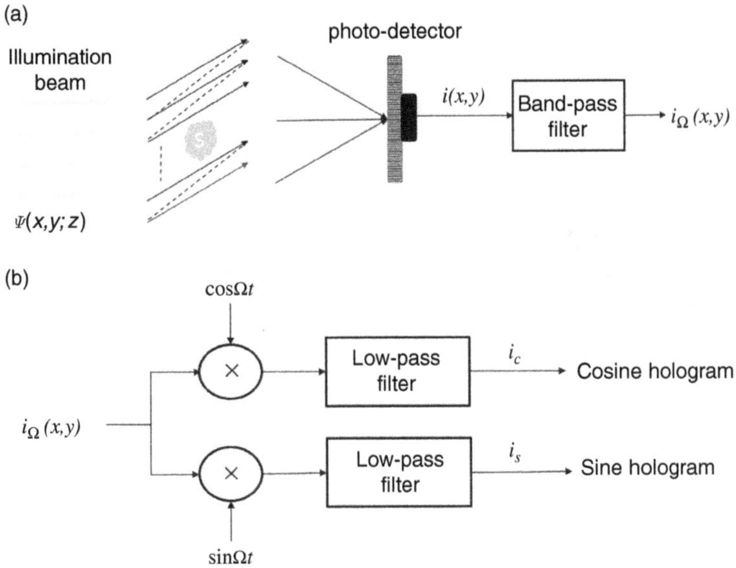

Figure 1.22 First (a) and second (b) stages of optical scanning holography (OSH).

$$i_s(x,y) = LPF[i_\Omega(x,y) \times \cos(\Omega t)] = K \sin\left[\frac{k_0}{2z}(x^2+y^2)\right] * |T(x,y;z)|^2, \qquad (1.33)$$

where $LPF[\ ]$ denotes a low-pass filtering operation. The pair of signals $i_c(x,y)$ and $i_s(x,y)$, which are referred to as the cosine and the sine hologram, are the real and the imaginary components of the hologram of the specimen, respectively.

1.7.4 Non-diffractive Optical Scanning Holography

The optical and electronic setups in a classical OSH, as described in Section 1.7.3, are rather complicated. The scanning beam is composed of a spherical wave and a plane wave that is modulated with a low-frequency sinusoidal wave at the heterodyne frequency. As the pair of beams have to be integrated into a single scanning beam, all the elements in the optical setups have to be positioned and aligned precisely. Subsequently, a pair of mixers are used to extract the cosine and sine holograms from the heterodyne signal of the photo-detector. Non-diffractive optical scanning holography (ND-OSH) [33,34] is a simplified version of OSH that can be realized with very few optical and electronics accessories. The ND-OSH system is shown in Figure 1.23. In ND-OSH, a pair of image patterns, $g_R(m,n;z)$ and $g_I(m,n;z)$, are projected from a commodity projector. The two image patterns are given by

$$g_R(m,n;z) = 1 + \cos\left\{\frac{2\pi}{\lambda}\left[\frac{(m\delta_d)^2 + (n\delta_d)^2}{2z}\right]\right\}, \qquad (1.34a)$$

and

$$g_I(m,n;z) = 1 + \sin\left\{\frac{2\pi}{\lambda}\left[\frac{(m\delta_d)^2 + (n\delta_d)^2}{2z}\right]\right\}. \tag{1.34b}$$

The specimen is sequentially scanned by the pair of image patterns along a zig-zag path, and a single pixel sensor is employed to record the reflected light. Scanning can be conducted by mounting the specimen on an X–Y platform that moves along a row-by-row and left-to-right path. The output signals of the single pixel sensor are two intensity images (each corresponding to one of the image patterns) representing the convolution between the 3-D image of the specimen and the pair of image patterns. Suppose the specimen is represented by a stack of D image planes $I_j(m,n)|_{1\leq\ j\leq\ D}$, each located at depth z_j (see Figure 1.14). Scanning of each depth plane results in two signals:

$$\begin{aligned} h_R(m,n;z_j) &= I_j(m,n) * g_R(m,n;z_j) \\ &= I_j(m,n) * \cos\left\{\frac{2\pi}{\lambda}\left[\frac{(m\delta_d)^2 + (n\delta_d)^2}{2z_j}\right]\right\} + [I_j(m,n) * 1] \\ &= I_j(m,n) * \cos\left\{\frac{2\pi}{\lambda}\left[\frac{(m\delta_d)^2 + (n\delta_d)^2}{2z_j}\right]\right\} + C, \end{aligned} \tag{1.35a}$$

and

$$\begin{aligned} h_I(m,n;z_j) &= I_j(m,n;z_j) * g_I(m,n;z_j) \\ &= I_j(m,n) * \sin\left\{\frac{2\pi}{\lambda}\left[\frac{(m\delta_d)^2 + (n\delta_d)^2}{2z_j}\right]\right\} + [I_j(m,n) * 1] \\ &= I_j(m,n) * \sin\left\{\frac{2\pi}{\lambda}\left[\frac{(m\delta_d)^2 + (n\delta_d)^2}{2z_j}\right]\right\} + C. \end{aligned} \tag{1.35b}$$

In Eqs. (1.35a) and (1.35b), $h_R(m,n;z_j)$ and $h_I(m,n;z_j)$ can be identified as the real and imaginary components of the hologram of the image plane $I_j(m,n)$, added with a constant term C. The hologram of the entire specimen can be obtained from these two components, as given by

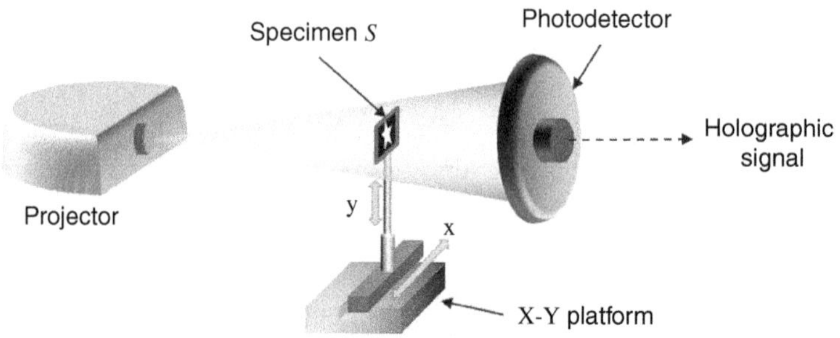

Figure 1.23 Non-diffractive optical scanning holography (ND-OSH) system.

$$
\begin{aligned}
H(m,n) &= \sum_{j=1}^{D} \{[h_R(m,n;z_j) - C] + i[h_I(m,n;z_j) - C]\} \\
&= \left\{ \sum_{j=1}^{D} [h_R(m,n;z_j) + ih_I(m,n;z_j)] \right\} - (1+i) \times C',
\end{aligned}
\tag{1.36}
$$

where $C' = D \times C$ is the overall intensity of the specimen, which can be obtained by exposing the specimen to a uniform white light and measuring the output of the photodetector. Similar to classical OSH, ND-OSH does not place rigid constraints on the coverage of the object scene and the size of the hologram.

1.8 MATLAB Simulation

This section includes the MATLAB source codes that are employed to conduct the simulations of the methods reported in this chapter. A summary of the simulation programs (each identified by the name of its main function call) and their correspondence to sections in the chapter are listed in Table 1.3.

The image "HK1" with the intensity distribution and depth map shown in Figures 1.12(a) and 1.12(b), respectively, is adopted in the simulation programs. The image is evenly divided into a left half-plane and a right half-plane, containing the characters H and K, respectively. The left half-plane is located at a distance of 0.07 m from the hologram plane, while the right half-plane is placed farther away at 0.08 m.

1.8.1 Simulation of Generating a Hologram with the Point-Based Method

The MATLAB code for this simulation is shown in Table 1.4. A digital Fresnel hologram of the double-depth image in Figure 1.12 is generated with the point-based method. The main function "Point_based" is called to load the source image file. Next, the function "PB_hologram" is called. Each object point is extracted from the source

Table 1.3 Summary of simulation programs in Chapter 1

Function name	Description of simulation	Associated sections
Point_based	Computer-generated hologram of a double-depth image with the point-based method	1.5.1
Off_axis	Simulation on capturing an off-axis hologram of a double-depth image	1.7.1
PSH	Simulation on capturing a hologram with the phase-shifting holography method	1.7.2

Table 1.4 MATLAB code of the point-based method for generating a digital Fresnel hologram of a set of object points

```
function Point_based()
    close all
    clc
    warning off
    fname='HK1';
    I=imread(strcat(fname,'.jpg'));    %Filename of intensity image
    I=im2double(I);
    psize=6.4*10^-6;                   %Pixel size of hologram=6.4 µm
    lamda=520*10^-9;                   %Wavelength=520 nm
    z1=0.07;                           %Depth of left half-plane of image
    z2=0.08;                           %Depth of right half-plane of image

    H=PB_hologram(I,psize,lamda,z1,z2);

    figure,imshow(mat2gray(real(H)));  %Display real part of hologram
    title('Real part of the digital Fresnel hologram');

    figure,imshow(mat2gray(imag(H)));  %Display imaginary part
                                       %of hologram
    title('Imaginary part of the digital Fresnel hologram');

    imwrite(mat2gray(real(H)),'Holo_PB_Real.jpg');
    imwrite(mat2gray(imag(H)),'Holo_PB_Imag.jpg');

    RES1=mat2gray(abs(Holo_recon(H,z1,psize,lamda)));
    RES2=mat2gray(abs(Holo_recon(H,z2,psize,lamda)));

    imwrite(RES1,'RES_PB_D1.jpg');
    imwrite(RES2,'RES_PB_D2.jpg');

    figure,imshow(RES1);               %Display reconstructed image at z1
    title('reconstructed image of the
  digital Fresnel hologram at z=z1');

    figure,imshow(RES2);               %Display reconstructed image at z2
    title('reconstructed image of the digital Fresnel hologram at z=z2');
end
function H=PB_hologram(I,psize,
  lamda,z1,z2)
    [N M]=size(I(:,:))
    wn=2*pi/lamda;                     %Wave number
    H=zeros(N,M);                      %Declare 2-D array for storing the
                                       %hologram

    %Compute and accumulate the object wave of each object point
    for row=1:N
     for col=1:M
       A=I(row,col);                   %Read intensity of a point at (row,
                                       %col)
       if A~=0                         %If intensity>0, it is an object
                                       %point

        if col<M/2
           z=z1;
        else
```

Table 1.4 (cont.)

```
            z=z2;
        end
        z_sq=z*z;                       %Square of the depth value
        prow=row-N/2-1;                 %Physical vertical position (row) of
                                          %the object point
        pcol=col-M/2-1;                 %Physical horizontal position (column)
                                          %of the object point

        %Declare meshgrid for calculating the object wave of a point source
        [kx ky]=meshgrid(-M/2-pcol:M/2-1-pcol,-N/2-prow:N/2-1-prow);
        kx=kx*psize;                    %Physical horizontal position
        ky=ky*psize;                    %Physical vertical position

        %Superimpose object wave of the object point to hologram
        H=H+A*exp(1i*wn*sqrt(kx.^2+ky.^2+z_sq));
        end
      end
   end
end

function RES=Holo_recon(H,z,psize,
  lamda)
   [N M]=size(H(:,:));                  %Size of hologram
   wn=2*pi/lamda;                       %Wave number

   %Generate Fresnel Zone Plate
   z2=z*z;                              %square of the depth value z
   [kx ky]=meshgrid(-M/2:M/2-1,-N/2:N/2-1);
   kx=kx*psize;                         %Physical horizontal position
   ky=ky*psize;                         %Physical vertical position
   fzp_c=exp(-1i*wn*sqrt(kx.^2+ky.      %Conjugate of FZP
   ^2+z2));
   %Compute reconstructed image
   RES=conv2(H,fzp_c);
end
```

image and its diffraction pattern, which is in the form of an FZP, is added to the hologram. Upon completion of the process, a digital Fresnel hologram of the double-depth image is generated. Subsequently, the hologram and its reconstructed images on the two focused planes are displayed and saved.

1.8.2 Simulation of Capturing an Off-Axis Hologram

This simulation, with the MATLAB code shown in Table 1.5, demonstrates capture of an off-axis hologram of a double-depth image. The main function "Off_axis_holo" is called to load the double image in Figure 1.12, and converts it into a digital Fresnel hologram with the function "Holo_gen." Next, the function "Off_Axis" is called to generate a plane wave that is inclined at an angle specified by the user. The inclined plane wave is added to the hologram, and the square of the magnitude of the hologram

Table 1.5 MATLAB code for generating an off-axis Fresnel hologram

```
function Off_axis_holo()
   close all
   clc
   warning off

   fname='HK1';
   I=imread(strcat(fname,'.jpg'));    %Filename of intensity image
   I=im2double(I);
   [N M]=size(I(:,:));                %Size of source image, identical to size
                                       %of hologram

   I1=I;                              %Image on first layer
   I1(:,M/2+1:M)=0;                   %Only left half-plane of source image in
                                       %first layer
   I2=I;                              %Image on second layer
   I2(:,1:M/2)=0;                     %Only right half-plane of source
                                       %image in second layer

   psize=6.4*10^-6;                   %Pixel size of hologram=6.4 μm
   lamda=520*10^-9;                   %Wavelength=520 nm
   z1=0.07;                           %Depth of left half-plane of image
                                       %(first layer)
   z2=0.08;                           %Depth of right half-plane of image
                                       %(second layer)

   H1=Holo_gen(I1,z1,psize,lamda);    %Hologram of first layer
   H2=Holo_gen(I2,z2,psize,lamda);    %Hologram of second layer
   H=H1+H2;                           %Hologram=sum of first and second layers
                                       %holograms

   %=========On-axis to off-axis conversion==================
   theta=input('Enter off-axis angle');
   HA=Off_Axis(H,psize,lamda,theta);

   %============Hologram reconstruction=====================
   figure,imshow(mat2gray(HA));       %Display reconstructed image at z0
   title('Off-axis hologram');
   imwrite(mat2gray(HA),strcat('Holo_OA_',num2str(theta),'.jpg'));

   RES1=mat2gray(abs(Holo_recon(HA,z1,psize,lamda)));
   RES2=mat2gray(abs(Holo_recon(HA,z2,psize,lamda)));
   imwrite(RES1,strcat('RES_OA_',num2str(theta),'_D1.jpg'));
   imwrite(RES2,strcat('RES_OA_',num2str(theta),'_D2.jpg'));

   figure,imshow(RES1);               %Display reconstructed image at z1
   title('reconstructed image of the digital Fresnel hologram at z=z1');
   figure,imshow(RES2);               %Display reconstructed image at z2
   title('reconstructed image of the digital Fresnel hologram at z=z2');
end
```

Table 1.5 (cont.)

```
function H=Holo_gen(I,z,psize,lamda)
    [N M]=size(I(:,:));
    wn=2*pi/lamda;                          %Wave number
    H=zeros(N,M);

    %Generate Fresnel Zone Plate
    z2=z*z;                                 %Square of the depth value z
    [kx ky]=meshgrid(-M/2:M/2-1,-N/2:N/2-1);
    kx=kx*psize;                            %Physical horizontal position
    ky=ky*psize;                            %Physical vertical position
    fzp=exp(1i*wn*sqrt(kx.^2+ky.^2+z2));

    %Generate hologram
    H0=conv2(fzp,I);
    [N0 M0]=size(H0(:,:));

    %Crop the hologram to its original size H_DIM
    H=H0((N0-N)/2:(N0+N)/2-1,(M0-M)/2:(M0+M)/2-1);
end

function HA=Off_Axis(H,psize,lamda,theta);
    theta=theta*pi/180;                     %Convert degree to radians
    M=max(abs(H(:)));
    H=H/M;
    [N M]=size(H(:,:))

    [kx ky]=meshgrid(-N/2:N/2-1,-M/2:M/2-1);
    R=exp(1i*2*pi.                          %Off-axis plane wave
  *ky*psize*sin(theta)/lamda);
    HA=abs(H+R).^2;                         %Add off-axis plane wave to hologram
end

function RES=Holo_recon(H,z,psize,
  lamda)
    [N M]=size(H(:,:));                     %Size of input hologram

    %Reconstruction area = Four times hologram size for accommodating the off-
  %axis reconstructed image
    %Move the original hologram to the
  %center of a larger hologram
    N4=N*4;M4=M*4;
    HA4=zeros(N4,M4);
    HA4((N4-N)/2:(N4+N)/2-1,(M4-M)/2:(M4+M)/2-1)=H;

    %Generate Fresnel zone plate
    wn=2*pi/lamda;                          %Wave number
    z2=z*z;                                 %Square of the depth value z
    [kx ky]=meshgrid(-M4/2:M4/2-1,-N4/2:N4/2-1);
    kx=kx*psize;                            %Physical horizontal position
    ky=ky*psize;                            %Physical vertical position
```

Table 1.5 (cont.)

```
    r=sqrt(kx.^2+ky.^2+z2);
    fzp_c=exp(-1i*wn*r)./r;             %Conjugate of FZP
    IR=conv2(HA4,fzp_c);
    [NR MR]=size(IR(:,:));

    %Crop to include only the region containing the reconstructed image
    N=round(N*1.5);M=round(M*1.5);
    RES=IR(NR/2-N:NR/2+N-1,MR/2-M:MR/
  2+M-1);
end
```

is taken to be the off-axis hologram. Subsequently, the reconstructed images at the two focused planes are computed, displayed, and saved.

1.8.3 Simulation of Capturing a Digital Fresnel Hologram with Four-Step Phase-Shifting Holography

Table 1.6 shows the MATLAB code for simulating the capture of a digital Fresnel hologram with four-step PSH. The main function "PSH," is called to load the double-depth image shown in Figure 1.12. The function "Holo_gen" is called to generate a complex-valued hologram of the double-depth image. Four interferograms are generated by mixing each hologram with one of the four phase-shifted versions of the reference plane wave. The four interferograms are then combined into a digital Fresnel hologram. Subsequently the reconstructed images at the two focused planes are computed, displayed, and saved.

1.9 Summary

In Sections 1.1–1.4, the fundamental principles of optical holography for capturing the optical waves of physical objects, and its difference from photography, are described. An image taken with photography, commonly referred to as a photograph, can only record a single view of an object scene. A hologram is capable of capturing the entire optical wavefront that impinged on it. As such, there should be little difference between observing a hologram and the physical object scene.

Numerical generation of a digital hologram, a technology that is commonly known as computer-generated holography, is presented in Section 1.5. Two important approaches in CGH – the point-based and the layer-based methods – are described. From the computational efficiency point of view, the point-based method is suitable for generating holograms of simple objects that comprise a small number of object points. However, for an object scene with large number of object points concentrated in a few depth planes, the layer-based method is preferred. The method for recovering a 3-D scene image from a digital hologram is provided in Section 1.6.

Table 1.6 MATLAB code simulating hologram acquisition with four-step PSH

```
function PSH()
   close all
   clc

   warning off
   fname='HK1';
   I=imread(strcat(fname,'.jpg'));    %Filename of intensity image
   I=im2double(I);
   [N M]=size(I(:,:));                 %Size of source image identical to size
                                         %of hologram

   I1=I;                               %First layer
   I1(:,M/2+1:M)=0;                    %Only left half-plane of source image in
                                         %first layer
   I2=I;                               %Second layer
   I2(:,1:M/2)=0;                      %Only right half-plane of source image
                                         %in second layer
   psize=6.4*10^-6;                    %Pixel size of hologram=6.4 µm
   lamda=540*10^-9;                    %Wavelength=540 nm

  %Generate object wave of image on first depth plane with the correlation-based
 %method
   z1=0.07;
   H=Holo_gen(I1,z1,psize,lamda);

   %Generate object wave of image on second depth plane with the correlation-
 %based method
   z2=0.08;
   H=H+Holo_gen(I2,z2,psize,lamda);
   MAX=max(abs(H(:)));
   H=H/MAX;

   IN1=abs(H+exp(1i*0)).^2;
   IN2=abs(H+exp(1i*pi/2)).^2;
   IN3=abs(H+exp(1i*pi)).^2;
   IN4=abs(H+exp(1i*3*pi/2)).^2;

   figure,imshow(mat2gray(IN1));       %Display interferogram 1
   title('1st interferogram');
   figure,imshow(mat2gray(IN2));       %Display interferogram 2
   title('2nd interferogram');
   figure,imshow(mat2gray(IN3));       %Display interferogram 3
   title('3rd interferogram');
   figure,imshow(mat2gray(IN4));       %Display interferogram 4
   title('4th interferogram');

   imwrite(mat2gray(IN1),'Holo_PSH_1.jpg');
   imwrite(mat2gray(IN2),'Holo_PSH_2.jpg');
   imwrite(mat2gray(IN3),'Holo_PSH_3.jpg');
   imwrite(mat2gray(IN4),'Holo_PSH_4.jpg');
```

Table 1.6 (cont.)

```
    %Compute real and imaginary components of the hologram from the four
   %interferograms
    HR=(IN1-IN3)/4;
    HI=(IN2-IN4)/4;

    %PSH hologram
    G=(HR+1i*HI);

    %Reconstruct images at the two focused planes from the PSH hologram
    RES1=abs(Holo_recon(G,z1,psize,lamda));
    figure,imshow(mat2gray(RES1));
    title('Reconstructed image of PSH hologram at depth = z1');
    RES2=abs(Holo_recon(G,z2,psize,lamda));
    figure,imshow(mat2gray(RES2));
    title('Reconstructed image of PSH hologram at depth = z2');

    imwrite(mat2gray(RES1),'RES_PSH_D1.jpg');
    imwrite(mat2gray(RES2),'RES_PSH_D2.jpg');
end

function H=Holo_gen(I,z,psize,lamda)
    [N M]=size(I(:,:));
    wn=2*pi/lamda;                          %Wave number
    H=zeros(N,M);

    %Generate Fresnel zone plate
    z2=z*z;                                 %Square of the depth value z
    [kx ky]=meshgrid(-M/2:M/2-1,-N/2:N/2-1);
    kx=kx*psize;                            %Physical horizontal position
    ky=ky*psize;                            %Physical vertical position
    fzp=exp(1i*wn*sqrt(kx.^2+ky.^2+z2));

    %Generate hologram
    H0=conv2(fzp,I);
    [N0 M0]=size(H0(:,:));

    %Crop the hologram to its original size H_DIM
    H=H0((N0-N)/2:(N0+N)/2-1,(M0-M)/2:(M0+M)/2-1);
end

function RES=Holo_recon(H,z,psize,lamda)
    [N M]=size(H(:,:));                     %Size of input hologram
    wn=2*pi/lamda;                          %Wave number

    %Generate Fresnel zone plate
    z2=z*z;                                 %Square of the depth value z
    [kx ky]=meshgrid(-M/2:M/2-1,-N/2:N/2-1);
    kx=kx*psize;                            %Physical horizontal position
    ky=ky*psize;                            %Physical vertical position
    r=sqrt(kx.^2+ky.^2+z2);
    fzp_c=exp(-1i*wn*r)./r;                 %Conjugate of FZP
```

Table 1.6 (cont.)

```
    IR=conv2(H,fzp_c);
    [NR MR]=size(IR(:,:));

    %Crop to include only the region containing the reconstructed image
    RES=IR((NR-N)/2:(NR+N)/2-1,(MR-M)/2:(MR+M)/2-1);
end
```

Approaching the end of this chapter, in Section 1.7 three different methods for capturing digital holograms of the physical object scene are described. The first method is similar to the art of optical holography, but instead of a photographic film, a digital camera is used to record the holographic waves emitted from the scene. As a digital camera can only record intensity information, the method can only be employed to capture an off-axis, amplitude-only hologram. The other two methods, known as phase-shifting holography and optical scanning holography, are capable of capturing both the magnitude and phase components of the holographic signals. The PSH is faster in operation, while the OSH can be used to capture holograms of large objects. Finally, a simplified version of OSH, known as non-diffractive optical scanning holography, is presented. ND-OSH is similar in principles to OSH, but the complexity of the optical and electronic setups is reduced.

Exercises

E.1.1 Referring to Figures 1.5(a) and 1.5(b), a pinhole camera is used to cast the image of two points, A and B, onto an image plane that is located at an axial distance of 0.03 m from the pinhole. The horizontal and vertical coordinates (in meters) of A, B and the pinhole are (0.01, 0.0), (0.005, −0.003), and (0.0, 0.0), respectively. Determine the positions of A', B' on the image plane.

E.1.2 The expression of a Fresnel zone plate (FZP) is given in Eq. (1.12). Given $\lambda = 0.52$ μm, $\delta_d = 6.4$ μm, and H_DIM = 2048, write a MATLAB program to display the magnitude and phase components of an FZP at different values of z_p, and explain why the distance z_p has to be sufficiently large in the generation of a digital Fresnel hologram.

E.1.3 Based on the optical settings in the previous question, perform a simulation with MATLAB to compare the computation time between generating a hologram of the single-depth image in Figure 1.12 with (a) the point-based method, and (b) the layer-based method. Repeat the above simulation with images comprising fewer object points. From the results, evaluate the pros and cons of these two methods in the generation of digital Fresnel holograms.

E.1.4 Extend the MATLAB codes in Section 1.8.2 to generate off-axis holograms of the double-depth image in Figure 1.12, based on a reference wave that is (a) inclined at an angle of 1.2° along the horizontal direction, and (b) inclined at an angle of 1.2° along both the horizontal and the vertical directions.

E.1.5 Modify the MATLAB codes in Section 1.8.3 to conduct a simulation of the four-step parallel phase-shifting holography method based on the phase-shifting array in Figure 1.21(b) for acquiring a digital Fresnel hologram of the double-depth image in Figure 1.12.

E.1.6 Perform a literature study on references [20] and [21]. Subsequently conduct a MATLAB simulation on the three-step and the two-step parallel phase-shifting holography method for acquiring a digital Fresnel hologram of the double-depth image in Figure 1.12.

References

[1] D. Gabor, "Microscopy by reconstructed wave-fronts," *Proc. Royal Soc. London (A), Math. Phy. Sci.*, vol. 197, no. 1051, pp. 454–487 (1949).

[2] W.L. Bragg, "Microscopy by reconstructed wave-fronts," *Nature*, vol. 166, pp. 399–400 (1950).

[3] P. Hariharan, *Optical Holography*, 2nd edition (Cambridge University Press, 1996).

[4] J. Goodman, *Introduction to Fourier Optics*, 3rd edition (Roberts and Company Publishers, 2004).

[5] Y.N. Denisyuk, "On the reflection of optical properties of an object in a wave field of light scattered by it," *Doklady Akademii Nauk SSSR*, vol. 144, no. 6, pp. 1275–1278 (1962).

[6] E. Leith and J. Upatnieks, "Reconstructed wavefronts and communication theory," *J. Opt. Soc. Am.*, vol. 52, pp. 1123–1130 (1962).

[7] J. Waters, "Three-dimensional Fourier-transform method for synthesizing binary holograms," *J. Opt. Soc. Am.*, vol. 58, pp. 1284–1288 (1968).

[8] B.R. Brown and A.W. Lohmann, "Complex spatial filtering with binary mask," *Appl. Opt.*, vol. 5, pp. 967–969 (1966).

[9] A.W. Lohmann and D.P. Paris, "Binary Fraunhofer holograms, generated by computer," *Appl. Opt.*, vol. 6, pp. 1739–1748 (1967).

[10] B.R. Brown and A.W. Lohmann, "Computer-generated binary holograms," *IBM J. Res. Dev.*, vol. 13, pp. 160–168 (1969).

[11] Y. Pan, Y. Wang, J. Liu, X. Li, and J. Jia, "Fast polygon-based method for calculating computer-generated holograms in three-dimensional display," *Appl. Opt.*, vol. 52, pp. A290–A299 (2013).

[12] J. Park, S. Kim, H. Yeom, H. Kim, H. Zhang, B. Li, Y. Ji, S. Kim, and S. Ko, "Continuous shading and its fast update in fully analytic triangular-mesh-based computer generated hologram," *Opt. Express*, vol. 23, pp. 33893–33901 (2015).

[13] H. Nishi and K. Matsushima, "Rendering of specular curved objects in polygon-based computer holography," *Appl. Opt.*, vol. 56, pp. F37–F44 (2017).

[14] J.P. Waters. "Holographic image synthesis utilizing theoretical methods," *Appl. Phys. Lett.*, vol. 9, no. 11, pp. 405–407 (1968).

[15] J. Chen and D. Chu, "Improved layer-based method for rapid hologram generation and real-time interactiveholographic display applications," *Opt. Express*, vol. 23, pp. 18143–18155 (2015).

[16] J.-S. Chen, Q. Smithwick, and D. Chu, "Implementation of shading effect for reconstruction of smooth layer-based 3D holographic images," Proc. SPIE 8648, 86480 R (2013).

[17] J.-S. Chen, D. Chu, and Q. Smithwick, "Rapid hologram generation utilizing layer-based approach and graphic rendering for realistic three-dimensional image reconstruction by angular tiling," *J. Electron. Imaging*, vol. 23, no. 2, 023016 (2014).

[18] U. Schnars and W.P.O. Jüptner, "Digital recording and numerical reconstruction of holograms," *Meas. Sci. Tech.*, vol. 13, no. 9, p. R85 (2002).

[19] I. Yamaguchi and T. Zhang, "Phase-shifting digital holography," *Opt. Lett.*, vol. 22, pp. 1268–1270 (1997).

[20] Y. Awatsuji, A. Fujii, T. Kubota, and O. Matoba, "Parallel three-step phase-shifting digital holography," *Appl. Opt.*, vol. 45, pp. 2995–3002 (2006).

[21] Y. Awatsuji, T. Tahara, A. Kaneko, T. Koyama, K. Nishio, S. Ura, T. Kubota, and O. Matoba, "Parallel two-step phase-shifting digital holography," *Appl. Opt.*, vol. 47, pp. D183–D189 (2008).

[22] X. Meng, L. Cai, X. Xu, X. Yang, X. Shen, G. Dong, and Y. Wang, "Two-step phase-shifting interferometry and its application in image encryption," *Opt. Lett.*, vol. 31, pp. 1414–1416 (2006).

[23] J.-P. Liu and T. Poon, "Two-step-only quadrature phase-shifting digital holography," *Opt. Lett.* vol. 34, pp. 250–252 (2009).

[24] Y. Awatsuji, M. Sasada, and T. Kubota, "Parallel quasi-phase-shifting digital holography," *Appl. Phys. Lett.*, vol. 85, pp. 1069–1071 (2004).

[25] S. Jeon D.-H. Kim, N.-C. Park, Y.-P. Park, and K.-S. Park, "Selective interpolation method for two-step parallel phase-shifting digital holography," *IEEE Trans. Mag.*, vol. 50, no. 7 (2014).

[26] P. Xia, Y. Shimozato, T. Tahara, T. Kakue, Y. Awatsuji, K. Nishio, S. Ura, T. Kubota, and O. Matoba, "Image reconstruction algorithm for recovering high-frequency information in parallel phase-shifting digital holography," *Appl. Opt.*, vol. 52, pp. A210–A215 (2013).

[27] P. Xia, Y. Awatsuji, K. Nishio, S. Ura, and O. Matoba, "Parallel phase-shifting digital holography using spectral estimation technique," *Appl. Opt.*, vol. 53, pp. G123–G129 (2014).

[28] M. Fujii, "A4-sized parallel phase-shifting digital holography system," *J. Disp. Tech.*, vol. 10, no. 2, pp. 132–137 (2014).

[29] M. Fujii, T. Kakue, K. Ito, T. Tahara, Y. Shimozato, Y. Awatsuji, S. Ura, K. Nishio, T. Kubota, and O. Matoba, "Construction of a portable parallel phase-shifting digital holography system," *Opt. Eng.*, vol. 50, no. 9, 091304 (2011).

[30] M. Lin, K. Nitta, O. Matoba, and Y. Awatsuji, "Parallel phase-shifting digital holography with adaptive function using phase-mode spatial light modulator," *Appl. Opt.*, vol. 51, pp. 2633–2637 (2012).

[31] T.-C. Poon and A. Korpel, "Optical transfer function of an acousto-optic heterodyning image processor," *Opt. Lett.*, vol. 4, pp. 317–319 (1979).

[32] T.-C. Poon, *Optical Scanning Holography with MATLAB* (Springer, 2007).

[33] P.W.M. Tsang and H. Lam, "Holographic vision system based on non-diffractive optical scanning holography and deep learning," SPIE/COS Photonics Asia 2019, Hangzhou, China (2019).

[34] P.W.M. Tsang and T.-C. Poon, "Non-diffractive optical scanning holography for hologram acquisition," in *Digital Holography and Three-Dimensional Imaging 2019* (Optical Society of America, 2019).

2 Fast Methods for Computer-Generated Holography

2.1 Introduction

In Chapter 1, computer-generated holography (CGH) was briefly reviewed. Basically, CGH is a simulation of optical holography whereby the interference pattern of light waves diffracted from the surface of a 3-D object is recorded on a light-sensitive medium known as a hologram. The process can be numerically computed by applying the Fresnel diffraction equation on a computer graphic model of the 3-D object. Although CGH is not a complicated process in itself, the amount of computation involved in generating a hologram is overwhelming, especially if the size of the hologram and the object space is large.

To estimate the number of arithmetic operations that are required to generate a hologram with the point-based method, the expression in Eq. (1.14) is rewritten:

$$H(m,n) = \sum_{p} A_p \times \exp\left[iw_n \times \sqrt{(m-m_p)^2 \times \delta_d^2 + (n-n_p)^2 \times \delta_d^2 + z_p^2}\right], \quad (2.1)$$

where $w_n = \frac{2\pi}{\lambda}$. Suppose a hologram is discretized into N rows and M columns, the total number of hologram pixels is $M \times N$. Assuming the constant δ_d^2 is known in advance, the total number of operations to compute a hologram is in the order of $(M \times N \times P)$, where P is the number of object points. Each operation is used to calculate the magnitude and phase of the optical wave of an object point that impinges on a hologram pixel. The arithmetic calculations involved in each operation are listed in Table 2.1.

As the pixel size of a hologram is in the order of wavelength of light, the number of pixels in a hologram is generally extremely large. For example, if the pixel size δ_d is 5 μm, a small hologram of 10 × 10 mm already contains two megapixels (2×10^6), which is equivalent to the number of pixels in a domestic television. If the number of object points is the same as the number of hologram pixels (i.e., $P = M \times N$), the number of operations is (2×10^{12}). Even with modern computers, generating a hologram at a video rate of 25–30 frames per second is still an extremely challenging task.

Eq. (2.1) shows that the computation of each hologram pixel is independent of the others. A straightforward way of speeding up the hologram-generation process, therefore, is to employ a set of processors to work in parallel. Each processor will be

Table 2.1 List of steps and the associated functions for computing the value of a hologram pixel of a single object point

Arithmetic functions		Quantity
Addition	$+$	2
Subtraction	$-$	2
Multiplication	$\times$	4
Square	a^2	4
Square root	$\sqrt{}$	1
Exponential function	exp()	1

dedicated to the computation of a unique hologram pixel. This kind of parallel operation can be realized with graphical processing units (GPUs). However, the speed enhancement is rather limited, and fast processors also lead to other problems such as size, power consumption, and heat dissipation. Numerous research attempts have been made since the early 1990s to speed up the hologram-generation process with fast algorithms that can be realized with parallel processors.

In this chapter, a number of fast hologram-generation methods that have been developed in the past decades will be reviewed. Some of these algorithms reduce the steps (or arithmetic operations) involved in the numerical realization of the CGH equations, while others are based on simplifying the hologram-generation process. This chapter describes a number of fast algorithms for generating a hologram from a set of object points.

2.2 Realization of CGH with Fourier Transform

As mentioned in Chapter 1, if a 3-D object space is composed of a large number of object points and few layers, the layer-based method is more computationally efficient than the point-based method. In the layer-based method, the object space is decomposed into a sequence of uniformly spaced image planes that are parallel to the hologram plane. Each image plane is located at an axial distance z_j from the hologram, and the object points on it are represented with the intensity distribution $I_j(m,n)$. The distance z_j is commonly referred to as the depth, and the image planes as the depth planes. Within the jth depth plane, all the object points have identical depth z_j from the hologram, and the hologram $H_j(m,n)$ can be generated as

$$H_j(m,n) = \sum_{u_p=0}^{M-1}\sum_{v_p=0}^{N-1} I_j(u_p,v_p) \times fzp(m-u_p, n-v_p; z_j), \tag{2.2}$$

where $fzp()$ is the Fresnel zone plate. Equation (2.2) can be rewritten as a convolution operation:

$$H_j(m,n) = I_j(m,n) * fzp(m,n;z_j). \tag{2.3}$$

Equation (2.3), in turn, can be realized in the frequency space. Suppose $\widetilde{I_j}(\omega_m,\omega_n)$ and $\widetilde{fzp}(\omega_m,\omega_n;z_j)$ represent the Fourier transform of $I_j(m,n)$ and $fzp(m,n;z_j)$, respectively, the hologram can be computed as

$$\begin{aligned} H_j(m,n) &= \mathcal{F}^{-1}[\widetilde{I_j}(\omega_m,\omega_n)\widetilde{fzp}(\omega_m,\omega_n;z_j)] \\ &= \mathcal{F}^{-1}[\mathcal{F}[I_j(m,n)]\Im[fzp(m,n;z_j)]], \end{aligned} \tag{2.4}$$

where $\Im[]$ is the forward Fourier transform, and $\Im^{-1}[]$ is the inverse Fourier transform. The hologram of the entire object space is equal to the sum of the holographic signals from all the image planes.

From the software implementation point of view, Eq. (2.4) can be conducted more swiftly than the convolution operation in Eq. (2.3), as it only involves a multiplication operation for each point in the frequency space. Both the forward and the inverse Fourier transform can be realized with fast Fourier transform (FFT), which is more computationally efficient than the convolution operation.

By the same principle, the numerical reconstruction process in each focused plane can also be speeded up with FFT. The reconstructed image $I_{R;j}(m,n)$ at the jth focused plane can be obtained from the hologram as follows:

$$I_{R;j}(m,n) = H(m,n) * fzp^{\dagger}(m,n;z_j), \tag{2.5}$$

which can be rewritten as

$$I_{R;j}(m,n) = H(m,n) * fzp^{\dagger}(m,n;z_j) = \mathcal{F}^{-1}\{\mathcal{F}[H(m,n)] \times \mathcal{F}[fzp^{\dagger}(m,n;z_j)]\}. \tag{2.6}$$

From the software implementation point of view, realizing Eq. (2.6) with FFT is considerable faster than convolution. However, Eq. (2.4) also shows that the computation can be overwhelming with a large number of depth planes. In the rest of this chapter, some of the contemporary methods for fast generation of digital holograms will be explained.

2.3 Direct Look-Up Table Method

A common practice to decrease the computation time in software programming is to store in the computer memory precomputed results of some tedious functions that have a high chance of being repeatedly used in the future. Such an approach is known as the look-up-table (LUT). The use of LUTs for speeding up the CGH process can be traced back to the works of Lucente in the early 1990s [1]. From Eq. (2.1) and Table 2.1, it can be seen that lots of arithmetic calculations are involved in deriving the expression $A_p \times fzp(m-m_p,n-n_p;z_p)$. These calculations can be waived if the values corresponding to all the combinations of the six variables $\{A_p,m,n,m_p,n_p,z_p\}$ have been

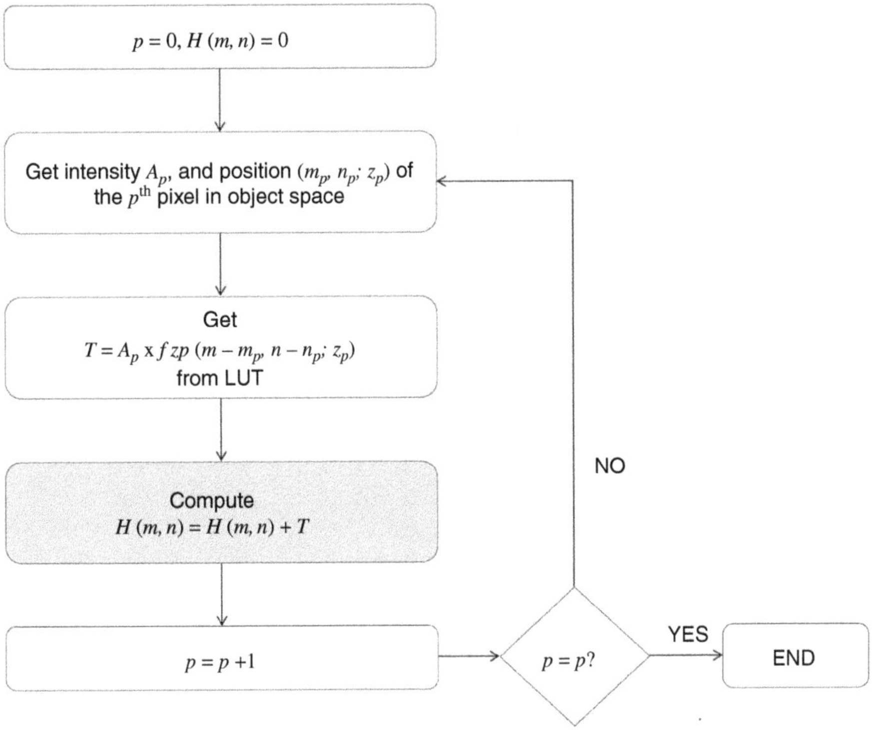

Figure 2.1 Flowchart for generating a hologram pixel with the LUT.

computed in advance and stored in the computer as a LUT. A hologram pixel at location (m, n) can then be obtained with the flowchart shown in Figure 2.1.

Assuming that the time taken to fetch the data and variables from the computer memory is negligible (which is generally valid for modern computers), generating a hologram pixel only requires one additional calculation (i.e., the shaded box in Figure 2.1) for each object point, which is significantly less than the computation listed in Table 2.1. For a hologram of size $M \times N$, and an object space with P object points, the number of arithmetic operations is reduced to $(M \times N \times P)$ additions. On the downside, a huge LUT is required to store the precomputed values. An estimation of the size of the LUT is given as follows. Suppose the object space is evenly partitioned into D image planes along the axial direction, and the intensity A_P is quantized into L levels. All the image planes are assumed to have identical horizontal and vertical extents as the hologram (i.e., size = $M \times N$). The total combinations from the six variables $\{A_p, m, n, m_p, n_p, z_j\}$ are $L \times M^2 \times N^2 \times D$. If each entry of the LUT is represented by B bits, the size of the LUT will be $(L \times M^2 \times N^2 \times D \times B)$ bits. As an example, a small square hologram (i.e., $M = N$) of size 1024×1024 representing an object scene of similar dimension, with eight depth planes, $L = 256$, and $B = 16$ (eight bits for each of the real and imaginary parts), will result in a LUT of $256 \times 1024^2 \times 1024^2 \times 8 \times 16 \approx 36\ 000$

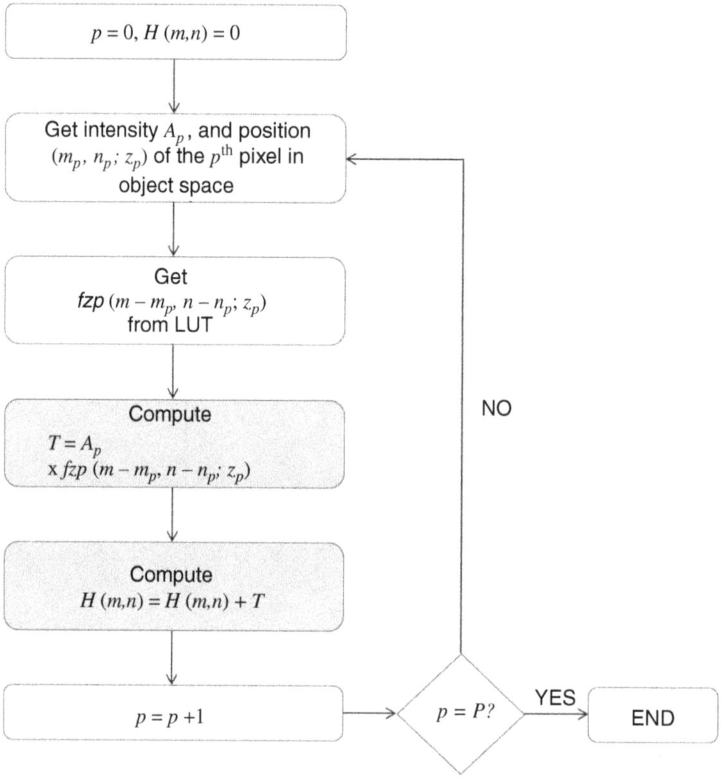

Figure 2.2 Flowchart for generating a hologram pixel with a smaller LUT.

terabits. Even with modern computers it is impractical, if not impossible, to set aside memory with size in the order of astronomical terabits for the LUT.

The memory problem can be alleviated with a hybrid approach, with some of the data being computed instead of storing in the LUT. A smaller LUT is constructed to store the values of the FZP function $fzp(m - m_p, n - n_p; z_p)$ for all combinations of values of the five variables $\{m, n, m_p, n_p, z_p\}$. The hologram pixel of an object point can be computed by multiplying its intensity A_p with the retrieved value from the LUT, as shown in the flowchart in Figure 2.2. The size of the LUT will be decreased dramatically from 35 840 terabits to around 140 terabits, at the expense of increasing the computation loading with $(M \times N \times P)$ multiplications. Even with the reduction in the size of the LUT, the memory requirement is still astronomical. In the subsequent subsections, different methods for trading off the size of the LUT and the speed enhancement in generating a hologram will be described.

2.4 Novel Look-Up Table Method

As mentioned in Section 2.3, the basic LUT method requires precomputation and storage of the FZP for all the combinations of the six variables $\{A_p, m, n, m_p, n_p, z_p\}$, resulting in a LUT of formidable size even for a small hologram. Although the size of the LUT can be reduced by discarding the intensity parameter A_p, the amount of computer memory is still enormous. The novel look-up table (N-LUT) method was proposed in [2], with the objective of further reducing the size of the LUT. A closer look at the function $fzp(m - m_p, n - n_p; z_p)$ reveals that it is simply a translation of the FZP $fzp(0, 0; z_p)$ of a point source located at an axial distance z_p from the hologram plane. Refer to Figure 1.14 on a layered partitioning of the object space. For each depth plane at z_j, a function $fzp(0, 0; z_j)$, which is referred to as the principal fringe pattern (PFP) is defined. Based on the PFP, the FZP of object points at other locations, $fzp(m - m_p, n - n_p; z_p)$, can be derived by translating the PFP. It is then sufficient to store the PFP for each depth plane in a small look-up table known as the N-LUT.

The construction of the N-LUT and its application in generating a hologram pixel from an object point are shown in Figure 2.3. Suppose the object space is divided into D depth planes within the range $[z_1, z_D]$, a set of PFP functions (PFP_1 to $PFPz_D$) is computed and stored in the N-LUT. To compute a hologram for a 3-D object, each depth plane is visited one by one. In each depth plane, the corresponding PFP is retrieved from the N-LUT. For an object point located at (m_p, n_p) on the jth depth plane and having intensity A_p, the PFP of the depth plane at z_j is shifted to the position (m_p, n_p). The translated PFP is then multiplied by A_p and accumulated on the hologram. This process is repeated for all the object points in the set of scan planes, resulting in a hologram of the 3-D object space.

Compared with the LUT method, the number of major arithmetic operations is slightly larger as the diffraction pattern of each object point is obtained by translating the PFP along the horizontal and vertical directions. Since the computation loading of these additional processes is negligible in practice, the overall complexity of the N-LUT method is about the same as that of the LUT method, comprising of $(M \times N \times P)$ multiplications and an equal number of addition operations. However, the size of the N-LUT is substantially reduced to $(M \times N \times D \times B)$ bits, which is

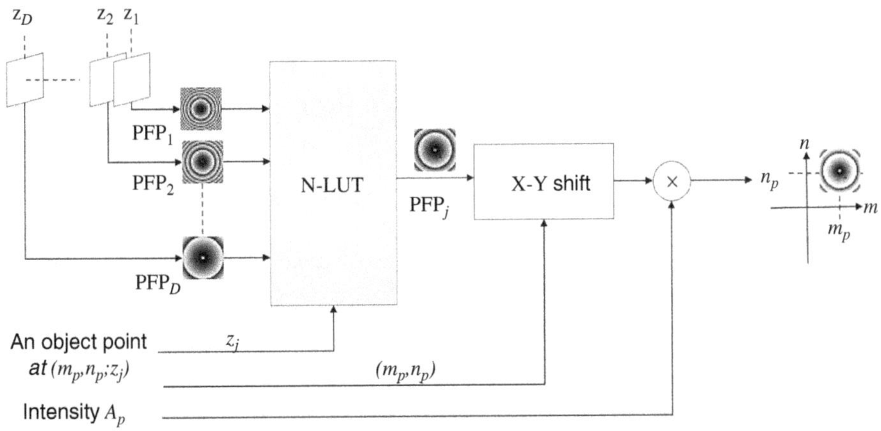

Figure 2.3 The N-LUT method.

$M \times N$ times smaller than that in the LUT. Referring to the previous example, the size of the N-LUT for $M = N = 1024$, $D = 8$, and $B = 16$ bits (2 bytes) is $1024 \times 1024 \times 8 \times 2 \approx 16.8$ megabytes. Reducing the size of the N-LUT by taking into account the pixel size and the reconstruction distance of the digital hologram was conducted and reported in [3].

2.5 The Line Scanning Method

Although the N-LUT is relatively smaller in size than the LUT, it still requires quite a lot of computer memory to store. To alleviate the memory requirement, Z. Yang has proposed the line scanning (LS) method [4] to further reduce the size of the N-LUT. As mentioned in Section 2.4, an N-LUT has to store a complete image of the PFP for each of the depth planes, resulting in a LUT of $(M \times N \times D \times B)$ bits. A PFP is an FZP in the form of concentric circular rings centered at the origin. Adopting polar coordinates, the location at (x, y) can be expressed as $x = r\cos\Theta$, $y = r\sin\Theta$, where r and Θ are the radius and azimuth of the hologram plane. The FZP at depth plane z_j can be rewritten in the polar form as

$$\begin{aligned} fzp(r, \Theta; z_j) &= \exp\left[iw_n\sqrt{(r\cos\Theta)^2 + (r\sin\Theta)^2 + z_j^2}\right] \\ &= \exp\left[iw_n\sqrt{r^2 + z_j^2}\right] = G(r; z_j). \end{aligned} \tag{2.7}$$

From the right-hand side of Eq. (2.7), it can be seen that the value of the function $fzp(r, \Theta; z_j)$ is only dependent on the radius. If the radius is uniformly discretized into R_N intervals, there will only be R_N different values for the 1-D function $G(r; z_j)$, with which the PFP on the corresponding depth plane can be generated. A LUT of size $(R_N \times D \times B)$ will be sufficient to encapsulate the PFPs of all the depth planes. Suppose the radius R_N of an FZP (which is half the width or height of a square hologram) is 512, the memory required to store the PFPs of a 3-D object scene with eight depth planes is $(R_N \times D \times B) = 512 \times 8 \times 2 \approx 8$ k bytes, which is about 2000 times smaller than that required to store the N-LUT.

In computing the hologram, a scan line is retrieved for each object point and used to reconstruct the PFP for $\Theta = [0, 2\pi)$. A straightforward implementation is to compute each point of the PFP as

$$fzp(m, n; z_j) = fzp(r\cos\Theta, r\sin\Theta; z_j). \tag{2.8}$$

From Eq. (2.8), it can be seen that each quadrant of the PFP is in fact a mirror image of the others, that is:

$$fzp(m, n; z_j)|_{m \leq 0; n \geq 0} = fzp(|m|, n; z_j), \tag{2.9}$$

and

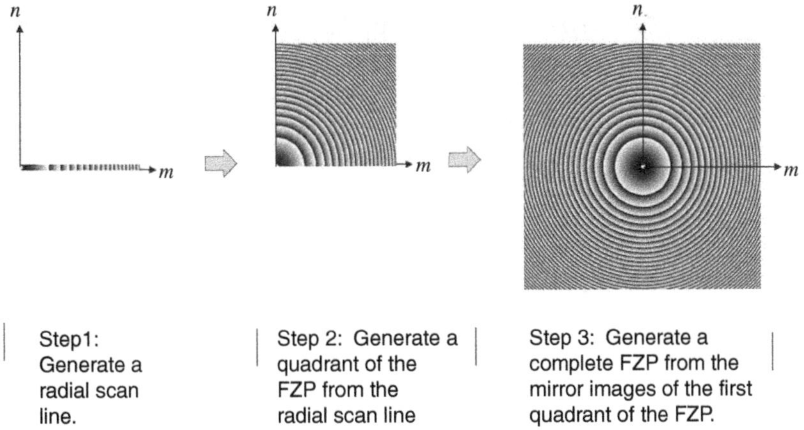

Figure 2.4 The LS method for generating the PFP.

$$fzp(m, n; z_j)|_{n \leq 0} = fzp(m, |n|; z_j). \tag{2.10}$$

Hence, computation of the PFP from the scan line can be simplified to the computation of the first quadrant of the PFP ($m; n \geq 0$), and the remaining quadrants are obtained from the symmetrical relation in Eqs. (2.9) and (2.10). The process can be divided into three steps, as illustrated in Figure 2.4. In step 1, a scan line is generated. In step 2, the first quadrant of the PFP is computed from the scan line. Finally, in step 3, the remaining quadrants of the PFP are duplicated from the mirror images of the first quadrant.

Expanding a radial line into a circular arc can be implemented with Bresenham's line algorithm [5]. However, some of the pixels in the PFP may be missing. An improved method has been proposed in [6], whereby the "arithmetical circle algorithm" [7] is used to generate the PFP from the scan line.

2.6 The Split-Look-Up-Table (S-LUT) Framework

Another method of reducing the size of the LUT, known as the split-look-up-table method (S-LUT), was proposed by Y. Pan et al. [8]. Previous LUT methods are more or less based on storing the FZP function in its original form. The S-LUT method employs an approximation of the FZP that is formed by the product of a pair of 1-D functions. The hologram is generated by accumulating the approximated FZP function of each object point. Referring to Eq. (2.1), the FZP of the pth point in the object space is a function of five variables $\{m, n, m_p, n_p, z_p\}$. By grouping the pairs of terms $\{m, m_p\}$ and $\{n, n_p\}$, the FZP can be simplified as a function of three variables

$$fzp(\Delta m, \Delta n; z_p) = \exp\left[\frac{i2\pi}{\lambda}\sqrt{\Delta m^2\delta_d^2 + \Delta n^2\delta_d^2 + z_p^2}\right], \tag{2.11}$$

where $\Delta m = (m - m_p)$ and $\Delta n = (n - n_p)$. The hologram $H(m, n)$ can be rewritten as

$$H(m,n) = \sum_p A_p \exp\left[i2\pi\lambda^{-1}\sqrt{\Delta m^2\delta_d^2 + \Delta n^2\delta_d^2 + z_p^2}\right]. \tag{2.12}$$

As $\Delta m^2 = |\Delta m|^2$ and $\Delta n^2 = |\Delta n|^2$,

$$H(m,n) = \sum_p A_p \exp\left[i2\pi\lambda^{-1}\sqrt{|\Delta m|^2\delta_d^2 + |\Delta n|^2\delta_d^2 + z_p^2}\right]. \tag{2.13}$$

Assuming $|\Delta m\delta_d| \ll z_p$ and $|\Delta n\delta_d| \ll z_p$, Eq. (2.13) can be approximated as

$$H(m,n) = \sum_p A_p \exp\left[i2\pi\lambda^{-1}\sqrt{|\Delta m|^2\delta_d^2 + z_p^2}\right]\exp\left[i2\pi\lambda^{-1}\sqrt{|\Delta n|^2\delta_d^2 + z_p^2}\right]. \tag{2.14}$$

Eq. (2.14) can be rewritten as

$$H(m,n) = \sum_p A_p O_H(|\Delta m|; z_p) O_V(|\Delta n|; z_p), \tag{2.15}$$

where $O_H(|\Delta m|; z_p) = \exp\left[i2\pi\lambda^{-1}\sqrt{|\Delta m|^2\delta_d^2 + z_p^2}\right]$ is the horizontal light modulation factor and $O_V(|\Delta n|; z_p) = \exp\left[i2\pi\lambda^{-1}\sqrt{|\Delta n|^2\delta_d^2 + z_p^2}\right]$ is the vertical light modulation factor. Both factors can be precomputed in advance and stored in the computer memory. In the generation of the hologram, the pair of functions are fetched from the memory based on values of $|\Delta m|$, $|\Delta n|$, and z_p. The remaining calculations will be the product of intensity A_p with the pair of retrieved functions for each object point, and the summing of the holographic signals for all the object points.

Without loss of generality, the pixel size of the object and hologram spaces are assumed to be identical, and both spaces have horizontal and vertical extents of M rows and N columns. The size of the LUTs for a single depth plane are M for $O_H(|\Delta m|; z_p)$ and N for $O_V(|\Delta n|; z_p)$ (where $0 \leq |\Delta m| < M$ and $0 \leq |\Delta n| < N$). However, since the computation of both light modulation factors is the same, only a single LUT is needed. Assuming a square hologram ($M = N$), this results in a LUT with M entries, which is much less than that required in the early LUT method. The size of the S-LUT for multiple plane will be multiplied by the number of depth planes that are used to discretize the 3-D object along the axial direction. The memory size of the S-LUT for each depth plane of a 1024×1024 hologram, with each entry of the S-LUT represented with 16 bits, is $1024 \times 16 \approx 16$ kbits.

Generation of a hologram pixel at location (m, n) for a single object point with the S-LUT method is summarized in Figure 2.5. Given an object point at position $(m_p, n_p; z_p)$, the horizontal and the vertical light modulation factors are retrieved from the LUTs. The pair of light modulation factors is then multiplied with the intensity A_p of the object point to generate the hologram pixel.

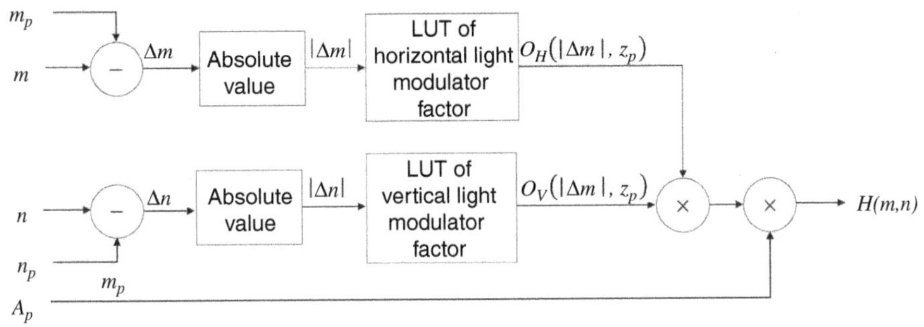

Figure 2.5 Computing a hologram pixel with the S-LUT method.

2.7 Compressed Look-Up-Table Method

The total size of the S-LUT for handling a 3-D object is equal to the number of entries of the S-LUT for a single depth plane, multiplied by the number of depth planes. This will lead to the use of a huge amount of computer memory if the number of depth planes is large. The compressed look-up-table (C-LUT) method [9] was developed by J. Jia et al. for reducing the size of the S-LUT at the expense of increasing the amount of computation, and adding extra restrictions on the depth range of the object space. The reconstructed image of a Fraunhofer hologram is similar to that of a Fresnel hologram, so that the hologram of a 3-D object comprising P object points can be derived from the Fraunhofer, instead of the Fresnel, diffraction as

$$H_F(m,n) = \sum_p A_p \exp\{i\pi\lambda^{-1}z_p^{-1}\delta_d^2[(m^2+n^2) - 2(m \times m_p + n \times n_p)]\}. \tag{2.16}$$

Assuming that the thickness of the 3-D object is considerably smaller than the mean distance z_μ between the object, Eq. (2.16) can be rewritten as

$$H_F(m,n) = \sum_p A_p L(m,n;z_p) O'_H(m,m_p) O'_V(n,n_p), \tag{2.17}$$

where $L(m,n;z_p) = \exp\left[i\pi\lambda^{-1}z_p^{-1}\delta_d^2(m^2+n^2)\right]$ is the longitudinal light modulation factor, $O'_H(m,m_p) = \exp\left[-i2\pi\lambda^{-1}z_\mu^{-1}\delta_d^2(m \times m_p)\right]$ is the horizontal light modulation factor, and $O'_V(n,n_p) = \exp\left[-i2\pi\lambda^{-1}z_\mu^{-1}\delta_d^2(n \times n_p)\right]$ is the vertical light modulation factor.

By replacing z_p with z_μ in the horizontal and vertical modulation factors, only the longitudinal light modulation factor is dependent on the distance of the depth plane. If $L(m,n;z_p)$ is obtained through computation instead of storing in the computer memory, a smaller LUT (which is referred to as the C-LUT), will be sufficient to store the pair of light modulation factors for all the combinations of values of (m,m_p) and

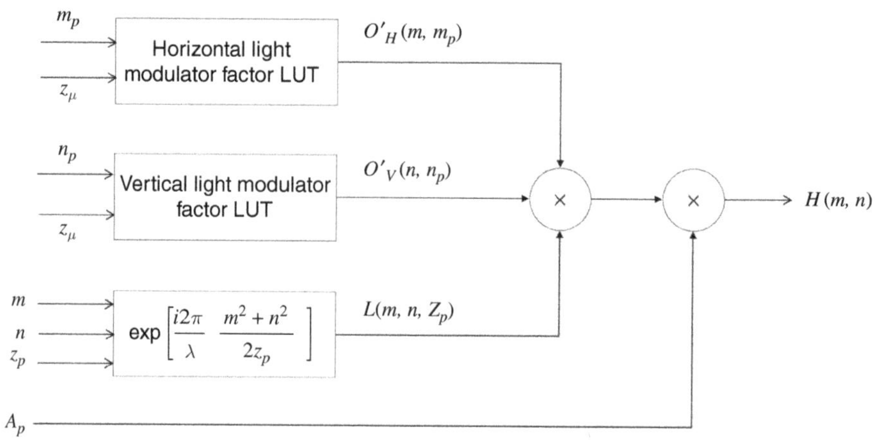

Figure 2.6 Computing a hologram pixel with the C-LUT method.

(n, n_p). For a square hologram, the overall size of the C-LUT is M^2. On the downside, the thickness of the object space has to be sufficiently small, and derivation of the longitudinal light modulation factor increases the computation loading as compared with the S-LUT method.

The process of applying the C-LUT method to generate a hologram pixel from an object point is shown in Figure 2.6. First, the horizontal and vertical light modulator factors for each object point are fetched from the computer memory. Next, the product of the pair of light modulator factors, together with the depth-dependent longitudinal light modulator factor and the intensity of the object point, are computed. The hologram is obtained from the sum of the diffracted waves of all the object points. Subsequent to [9], further reduction in the size of the C-LUT by two times has been achieved with the accurate compressed look-up-table (AC-LUT) method in [10].

2.8 Wavefront Recording Plane Method

The hologram-generation methods described in previous sections have assumed that the light wave of each object point is recorded on the entire hologram plane. For a single object point, its hologram will be an FZP that is bounded by the perimeter of the hologram. In practice, due to the finite resolution of a digital hologram, the diffraction angle is limited to $\theta_{max} = \sin^{-1}[\lambda/(2\delta_d)]$, where λ is the wavelength of light and δ_d is the pixel size of the digital hologram. Taking this into account, the hologram should only record the part of the FZP that is within the area covered by the maximum diffraction angle of the object point, as shown in Figure 2.7. Furthermore, the closer the distance between the hologram and the object point, the smaller will be the coverage of the light beam of the object point. This is in line with the notion of a virtual window in [11], where only a small region of the hologram is computed for each object point, resulting in a significant reduction of the computational load.

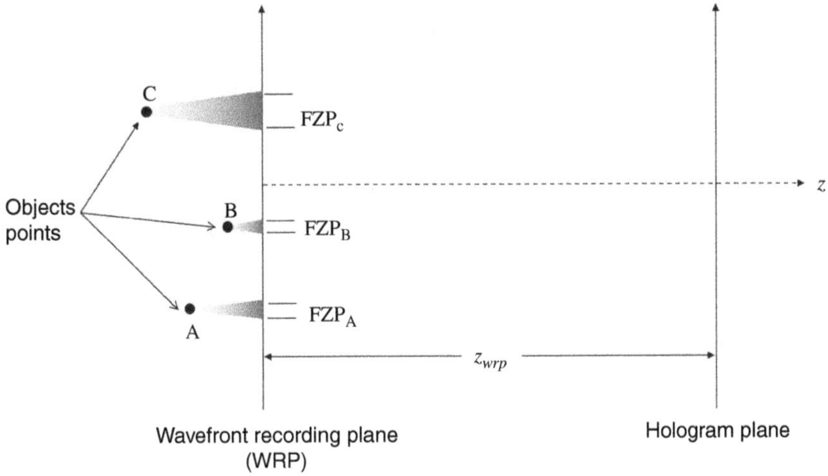

Figure 2.7 Spatial relation between the object space, the WRP, and the hologram.

A more formal adoption of the concept of a virtual window in hologram generation was proposed by T. Shimobaba et al. in [12]. A hypothetical plane known as the wavefront recording plane (WRP) is placed close to the object space. Each object point emits a cone of light wave with an aperture that is limited by the maximum diffraction angle θ_{max}. The directrix of the cone on the WRP is the perimeter within which the FZP of the object point needs to be computed. A side view of the hologram, the WRP, and three object points is shown in Figure 2.7. Each object point is projecting a small FZP on the WRP, and it can be seen that the size of the FZP is inversely proportional to the distance between an object point and the WRP. The WRP is in fact a hologram that is at near proximity to the object wave. After the diffractive waves of all the object points have been computed on the WRP, the latter is expanded to a full hologram on the hologram plane. The mechanism is similar to the generation of a H2 hologram, with the WRP being the first hologram, which is copied to a second hologram at a farther distance.

Details of the WRP method in generating a hologram can be divided into two major stages.

Stage 1

Referring to Figure 2.7, each object point casts a small FZP on the WRP plane that is bounded by the circular directrix of the cone of light wave. However, to simplify the software implementation, the circular region can be replaced with a square support. The side length of the square support, denoted by τ, is equal to the diameter of the directrix. The object wave, the support of the pth object point on the WRP, and the distance between the object point and a pixel at (m,n) on the WRP are denoted by $W(m,n)$, S_p, and $r_{m;n;p}$, respectively. The FZP within the support S_p corresponding to the pth object point is given by

$$w_p(m,n)|_{(m,n)\in S_p} = A_p r_{m;n;p}^{-1} \exp(i2\pi\lambda^{-1} r_{m;n;p}). \tag{2.18}$$

The object wave on the WRP is the sum of the FZPs of all the object points:

$$W(m,n) = \sum\nolimits_p w_p(m,n)|_{(m,n)\in S_p}. \tag{2.19}$$

As the support S_p is a small square, the computation involved in stage 1 is significantly lower than that required for computing the hologram directly with Eq. (2.1).

Stage 2

After the WRP is generated, it is projected onto the hologram with the Fresnel diffraction equation given by

$$H(m,n) = W(m,n) * h(m,n;z_{wrp}), \tag{2.20}$$

where z_{wrp} is the distance between the WRP and the hologram, and $h(m,n;z_{wrp})$ is the free space impulse response in Fourier optics. To further speed up the computation, the convolution operation in Eq. (2.20) can be implemented in the Fourier space as

$$H(m,n) = \mathcal{F}^{-1}\{\mathcal{F}[w(m,n)] \times \mathcal{F}[h(m,n;z_{wrp})]\}. \tag{2.21}$$

As the Fourier transform of $h(m,n;z_{wrp})$ can be precomputed, Eq. (2.21) can be implemented with a multiplication, a forward Fourier transform, and an inverse Fourier transform operation. All three operations can be swiftly conducted in less than 10 ms with a commodity GPU device. With the use of a LUT [13] for storing precomputed values of $w_p(m,n)$ for all combinations of $\{m,n,A_p,r_p\}$, a 2048×2048 hologram representing 3×10^4 object points can be generated with the WRP method at around 10 frames per second.

Simulation on generating a hologram with the WRP is conducted with MATLAB code in Section 2.11.1. As before, the object space is the double-depth image in Figure 1.12. The spatial relation between the double-depth image, the WRP, and the hologram is shown in Figure 2.8. From the figure, it can be seen that the object space is quite near to the WRP, with the left and the right sides of the image at axial distances of $z_0 = 0.01$ m and $z_1 = 0.02$ m from the WRP. The separation between the WRP and the hologram is $z_{wrp} = 0.06$ m. The length τ of the square support is set to 64.

After running the MATLAB codes, the real and the imaginary components of the object wave on the WRP are shown in Figures 2.9(a) and 2.9(b). The real and imaginary components of the hologram are shown in Figures 2.9(c) and 2.9(d), and the reconstructed images are shown in Figures 2.9(e) and 2.9(f). The time taken to generate the hologram with the WRP method is about five times less than the direct point-based method in Section 2.1, but the reconstructed image of each character on its focused plane is similar to the original ones in Figure 1.12.

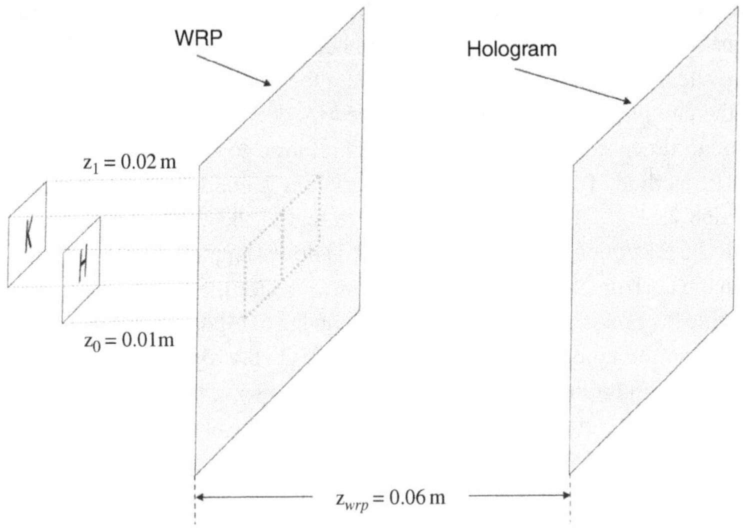

Figure 2.8 Spatial relation between the double-depth image, the WRP, and the hologram.

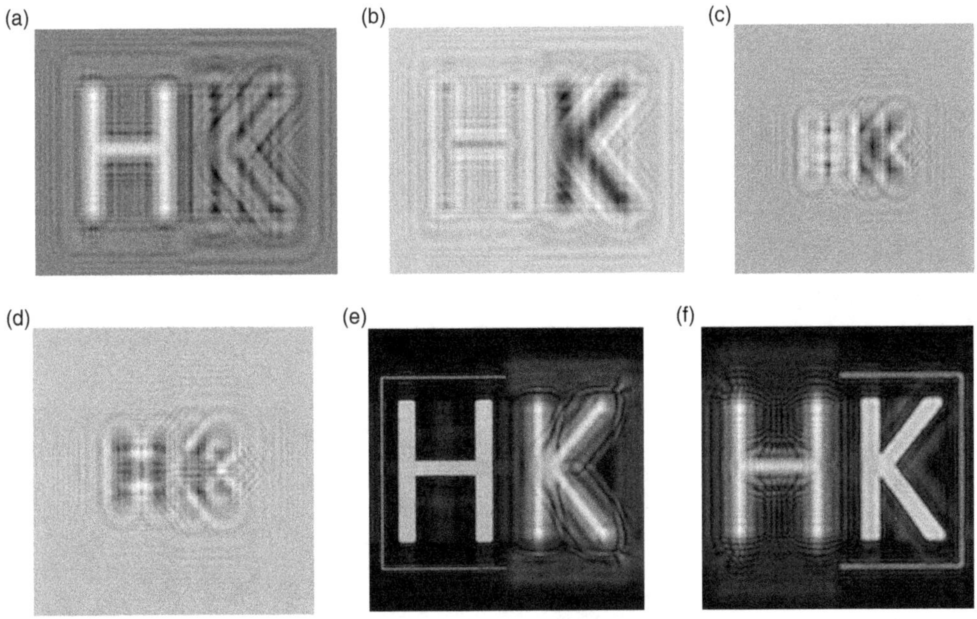

Figure 2.9 (a) Real component of the object wave on the WRP. (b) Imaginary component of the object wave on the WRP. (c) Real component of the hologram. (d) Imaginary component of the hologram. (e, f) Reconstructed images of the hologram, at depth planes 0.07 m and 0.08 m, respectively.

From Eqs. (2.18) and (2.19), it can be seen that the time taken to compute the WRP signal is directly proportional to the number of object points. As such, the time taken to generate a hologram can be rather lengthy for a dense object space. Further reduction of the computation time can be achieved by decreasing the number of object points through downsampling. This method is referred to as the downsampled WRP (DS-WRP) method. To illustrate this, a simulation is conducted using MATLAB code in Section 2.11.2. The program is similar to that of the WRP method in Section 2.11.1, with the exception that the object space is downsampled by a factor of DS = 8 along both the horizontal and vertical directions to reduce the number of object points. After running the program on the double-depth image in Figure 1.12, the object waves on the WRP and the hologram are shown in Figure 2.10(a–d). The reconstructed images of the hologram on the two focused planes are shown in Figures 2.10(e) and 2.10(f).

A fast WRP method that does not require the use of LUTs has been demonstrated in [14]. In this approach, a small FZP of each object point is computed with a fast algorithm, hence saving the large amount of computer memory that is used in storing the LUT. Another fast method has been proposed in [15,16], whereby the wavelet transform is employed to reduce the amount of computation. Later, the WRP method was extended to generate color holograms [17], as well as holograms that are smaller than the object images with the use of shifted-Fresnel diffraction [18].

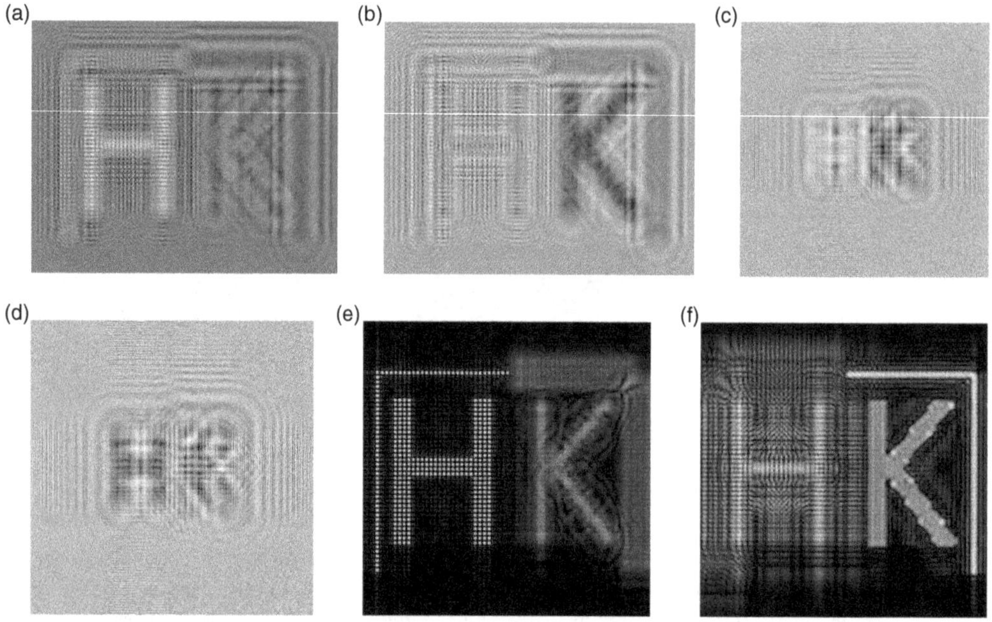

Figure 2.10 WRP method with downsampling the source image with DS = 8. (a) Real component of the object wave on the WRP. (b) Imaginary component of the object wave on the WRP. (c) Real component of the hologram. (d) Imaginary component of the hologram. (e, f) Reconstructed images of the hologram, at depth planes 0.07 m, and 0.08 m, respectively.

The WRP approach has two major disadvantages. First, in order to speed up the hologram-generation process, the number of object points should be reduced. This can be achieved by downsampling the intensity image of the object scene, resulting in a sparse image of degraded quality. Downsampling also imposes aliasing error on the image, especially regions with high spatial frequency content. Although aliasing error can be lowered by limiting the bandwidth of the image, the lines and edges pattern will be blurred. On the other hand, if the downsampling factor is small, more object points will be included and the computation time of hologram generation will be increased accordingly. The second problem is that the separation between the WRP must be close to the object points, so that the support S_p is small for all the object points. This is important for the WRP method as the computation time, as depicted in Eq. (2.18), is proportional to the size of the support. In another words, in order to maintain the computation efficiency of the WRP method, the thickness (depth range) of the object must be small. This problem has been addressed with the double WRP [19] and multiple WRP methods [20–22]. In these approaches, the object space is partitioned into two or more narrow zones along the axial direction. A WRP is assigned to each zone to record the diffraction patterns of its constituting object points. The hologram is generated by summing the optical waves from the WRPs of all the zones. Including more WRPs lengthens the time taken to generate the hologram, and hence degrades the computational efficiency. An alternative WRP method, known as the tilted WRP method [23], is also effective in generating holograms of deep structure object scenes.

2.9 Interpolated WRP Method

As mentioned in Section 2.8, the WRP method faces the dilemma of trading off image quality with computational efficiency. The interpolated WRP (IWRP) method [24] was developed to alleviate this problem. The concept of this method is to generate the diffracted pattern for a small region of points instead of individual ones. The process can be divided into three steps.

Step 1

The object space and the WRP are first uniformly divided into non-overlapping square blocks of size $\tau \times \tau$. For each block in the object space (which is indexed with the variable j), its pixels are assigned the mean amplitude $\mu_{a;j}$ and mean depth values $\mu_{d;j}$ (with reference to the WRP) of its constituting pixels. If $\mu_{a;j} > 0$, the partition is referred to as a "non-empty partition." An identical area (i.e., the same size and at the same horizontal and vertical positions) is partitioned on the WRP plane for each of the square blocks on the object space. The area bounded by the jth square block, which is identical for both the object space and the WRP, is referred to as the support, and denoted with the symbol S_j. It can be inferred that this mechanism is equivalent to downsampling, and subsequently interpolation of the object space. In downsampling, each partition is reduced to a single sample point carrying the mean amplitude and mean depth value of the object points in it. In interpolation, all the pixels within the partition are duplicated from the sample point. The diffraction fringe pattern $w_j(m,n)$ on the WRP corresponding to the jth block (i.e., support S_j) is given by

$$w_j(m,n)|_{(m,n)\in S_j} = \sum_p \frac{\mu_{a;j}}{r_{m;n;p}} \exp\left(\frac{i2\pi r_{m;n;p}}{\lambda}\right), \tag{2.22}$$

where $r_{m;n;p}$ is the distance between the pth object point in S_j, and a pixel at location (m,n) within the support S_j on the WRP. From the expression on the right-hand side of Eq. (2.22), $w_j(m,n)$ is in fact the hologram of the jth square block of pixels in the object space. Similar to the WRP method, a LUT can be constructed to store precomputed diffraction patterns of the small blocks for different combinations of $\mu_{a;j}$ and $\mu_{d;j}$.

Step 2

After the hologram fringe patterns of all the non-empty partitions have been derived with Eq. (2.22), the holographic image on the WRP will be formed by the union of the fringe pattern of individual partition as

$$W(m,n) = \bigcup_{j=0}^{T-1} w_j(m,n), \tag{2.23}$$

where T is the number of non-empty partitions. As the WRP now contains the fringe patterns of an interpolated object image, it is renamed as the interpolated WRP. At first sight, it seems that Eq. (2.23) is similar to Eq. (2.19). There is no arithmetic operations in Eq. (2.23) and a union of the fringe patterns, which only requires a negligible amount of computations, is conducted instead. With the IWRP approach, the generation of a 2048 × 2048 hologram at a video rate of 25 frames per second, comprising four million object points, is demonstrated in [24].

Step 3

The optical wavefront on the WRP is expanded to a hologram $H(m,n)$ that is parallel to it, and placed at a distance z_{wrp} before it. Denoting $\mathcal{F}$ and $\mathcal{F}^{-1}$ as the forward and inverse Fourier transform,

$$H(m,n) = \mathcal{F}^{-1}\{\mathcal{F}[W(m,n)] \times \mathcal{F}[h(m,n;z_{wrp})]\}. \tag{2.24}$$

2.10 The Warped WRP Method

The WRP can be considered as a hologram that is very close to the object space. Due to the short distance, the optical wave of each object point will only cover a small area on the WRP, as shown in Figure 2.11. Any modification made to the WRP therefore only affects a small area of the object space. If the modification is an enlargement of a region on the

WRP, it is equivalent to moving the object points (that are associated with that region) farther away from the WRP. The warped WRP (WWRP) method [25], which is applied in fast generation of holograms, is based on this principle. The upper part of Figure 2.11 shows an image plane containing a pair of characters, H and K, and a WRP that is at close proximity and parallel to the image plane. Both characters are equidistant to the WRP. Due to the short distance between the image plane and the WRP, the optical waves corresponding to each character on the WRP are roughly confined within the area enclosing the character (i.e., the dotted boundary on the WRP). If the diffraction pattern of the character H (i.e., the region highlighted in gray) is magnified, it will be equivalent to moving the character farther away from the WRP, as shown in the lower half of Figure 2.11.

The WWRP method can be divided into four stages, as depicted in Figure 2.12. The object scene is a 3-D surface represented by an intensity image $I(m,n)$. The object space is partitioned into J regions, each having its own depth from the hologram. Let $S_j(m,n)$ and z_j denote the pixel locations of the jth region, and the depth value of the region, respectively. In the first stage, each region is scaled by a factor a_j, resulting in a new intensity image $I'(m,n)$. As will be explained later, the scaling factor a_j is determined by the depth of its corresponding region. In the second stage, a WRP hologram of $I'(m,n)$ is computed. The WRP is spatially partitioned into J regions, each corresponding to the pixel locations in $S_j(m,n)$. In the next stage, each area in the WRP is scaled by a factor $1/a_j$. In the final stage, the WRP hologram is converted into a hologram.

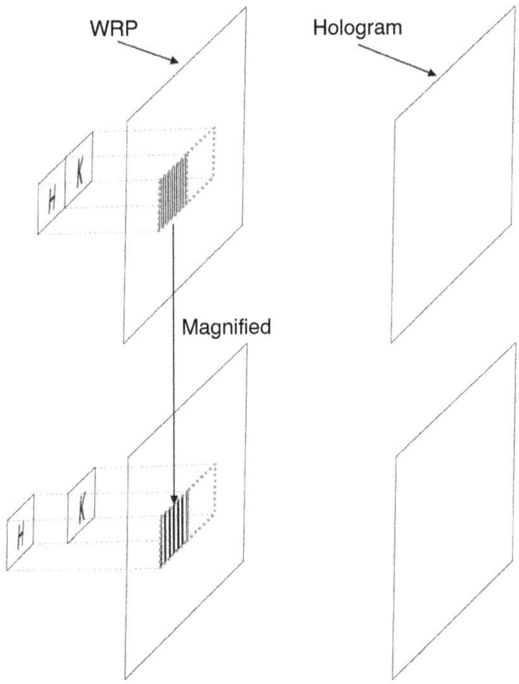

Figure 2.11 Concept of the WWRP method.

For simplicity, refer to the object space of the double-depth image shown in Figure 2.11. In this example, the object space is composed of a left region S_0 containing the character H, and right region S_1 containing the character K. The principle can be extended to an object scene with over two image planes.

Stage 1

In the first stage, each region S_j in the intensity image is scaled by a factor a_j along the horizontal and vertical directions. Determination of the scaling factor a_j will be discussed later. The image after scaling is given by

$$I'(m,n)|_{(m,n)\in S_j} = I\left(\frac{m}{a_j},\frac{n}{a_j}\right). \tag{2.25}$$

In the example of Figure 2.11, S_0 and S_1 are located at distances $z_0' = 0.02$ m and $z_1' = 0.01$ m from the WRP. The prewarped intensity image is assumed to be on the image plane S_1, which is closest to the WRP. Applying the Fresnel diffraction equation, the WRP hologram of the prewarped intensity image is generated as

$$w(m,n) = I'(m,n) * h(m,n;z_1'), \tag{2.26}$$

where $h(m,n;z_1') = \exp\left(i2\pi\lambda^{-1}\sqrt{(m\delta_d)^2 + (n\delta_d)^2 + (z_1')^2}\right)$ is the free space impulse response, which can be approximated as

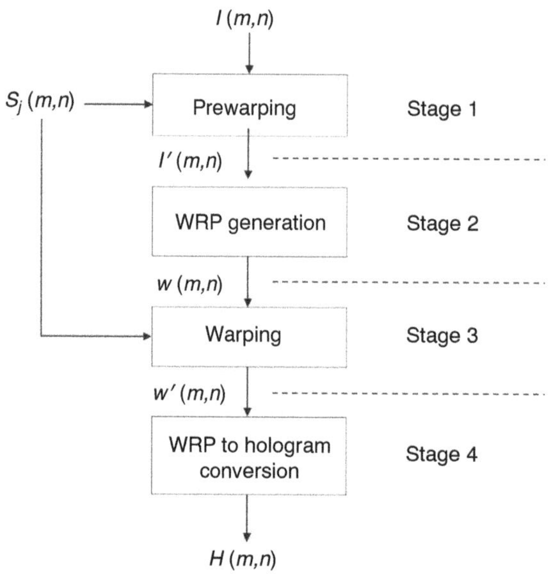

Figure 2.12 Flowchart of the WWRP method.

$$h(m,n;z_0') \approx \exp\left[\frac{i2\pi}{\lambda}\frac{(m^2\delta_d^2 + n^2\delta_d^2)}{2z_1'}\right]. \tag{2.27}$$

The WRP represents a planar image that is parallel to the hologram and the image $I'(m,n)$, hence all the pixels are equidistant to the hologram. Due to the close proximity of the WRP and the object space, the light wave cast by each region on the WRP is more or less confined within the area covered by the region. As such, the WRP can be considered as the union of WRP components each composed of the hologram fringes of individual regions:

$$w(m,n) = \bigcup_j w_j(m,n). \tag{2.28}$$

In the example shown in Figure 2.10, there are only two regions.

Stage 2

For stage 1 it has been assumed that all the regions of the image $I'(m,n)$ are at equal distance z_1' to the WRP. However, the depth value of each region S_j should be z_j' instead of z_1'. In stage 2, the area corresponding to the region S_j on the WRP is enlarged by a factor a_j, so that the actual depth of S_j can be recorded on the WRP. Applying the scaling factor $1/a_j$ to the WRP hologram $w(m,n)$,

$$\begin{aligned} w_j'(m,n) &= w_j(a_jm, a_jn) \\ &= I'(a_jm, a_jn)|_{m;n\in S_j} * \exp\left[\frac{i2\pi}{\lambda}\frac{\left[(a_jm)^2\delta_d^2 + (a_jn)^2\delta_d^2\right]}{2z_1'}\right] \\ &= I'(a_jm, a_jn)|_{m;n\in S_j} * \exp\left[\frac{i2\pi}{\lambda}\frac{(m^2\delta_d^2 + n^2\delta_d^2)}{2(z_1'/a_j^2)}\right]. \end{aligned} \tag{2.29}$$

Substituting Eq. (2.25) into Eq. (2.29), the jth region on the WRP is changed to

$$\begin{aligned} w_j'(m,n)|_{m;n\in S_j} &= w_j(a_jm, a_jn) \\ &= I(m,n)|_{m;n\in S_j} * \exp\left[\frac{i2\pi}{\lambda}\frac{(m^2\delta_d^2 + n^2\delta_d^2)}{2(z_1'/a_j^2)}\right] \\ &= I(m,n)|_{m;n\in S_j} * h(m,n;z_1'/a_j^2). \end{aligned} \tag{2.30}$$

Assigning $\frac{z_1'}{a_j^2} = z_j'$, Eq. (2.30) can be rewritten as

$$w_j'(m,n)|_{m;n\in S_j} = I(m,n)|_{m;n\in S_j} * h(m,n;z_j'). \tag{2.31}$$

It can be seen from Eqs. (2.30) and (2.31) that by rescaling the part of the WRP corresponding to S_j by a factor a_j^2, and setting $a_j = \sqrt{z_1'/z_j'}$, the part of the WRP corresponding to S_j will be representing the region relocated to depth z_j' from the WRP.

In the third stage, a new WRP hologram is composed from the union of the scaled WRP components:

$$w'(m,n) = \bigcup_j w_j'(m,n). \tag{2.32}$$

Subsequently, the WRP hologram is projected onto the hologram plane, resulting in a hologram representing the object scene comprising all the regions S_0 to S_{J-1}, each locating at its own depth from the hologram. The hologram $H(m,n)$ is given by

$$H(m,n) = w'(m,n) * h(m,n;z_{wrp}). \tag{2.33}$$

The MATLAB code in Section 2.11.3 is used to demonstrate the WWRP method. The source image $I(m,n)$, having a size of 256×256, is shown in Figure 2.13(a). The image is divided into four regions $[S_0, S_1, S_2, S_3]$, each corresponding to a different depth from the hologram, as shown in Figure 2.13(b). The WWRP method is then applied to generate the hologram for the image $I(m,n)$, with respect to the depth map in Figure 2.13(b). First Eq. (2.25) is applied to prewarp the image, resulting in a new image as shown in Figure 2.13(c). The size of each character is scaled down by a factor proportional to its distance from the hologram. The prewarped image is then converted into a digital Fresnel hologram on the WRP that is located at an axial distance of

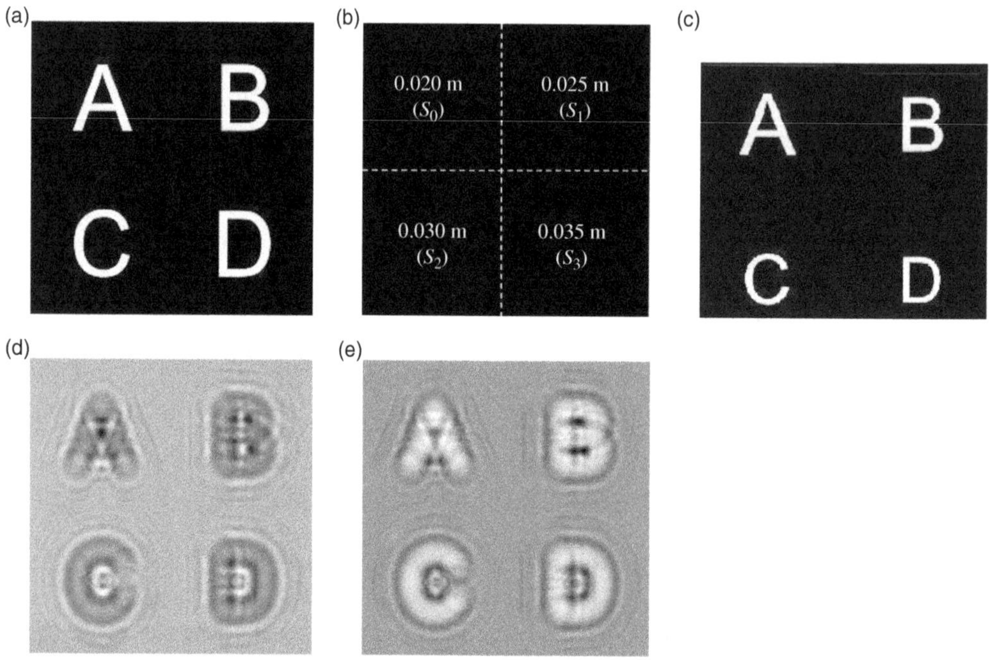

Figure 2.13 (a) Intensity of the source image. (b) Depth map of the source image. (c) Prewarped image. (d) Real component of the WRP hologram of the prewarped image. (e) Imaginary component of WRP hologram of the prewarped image.

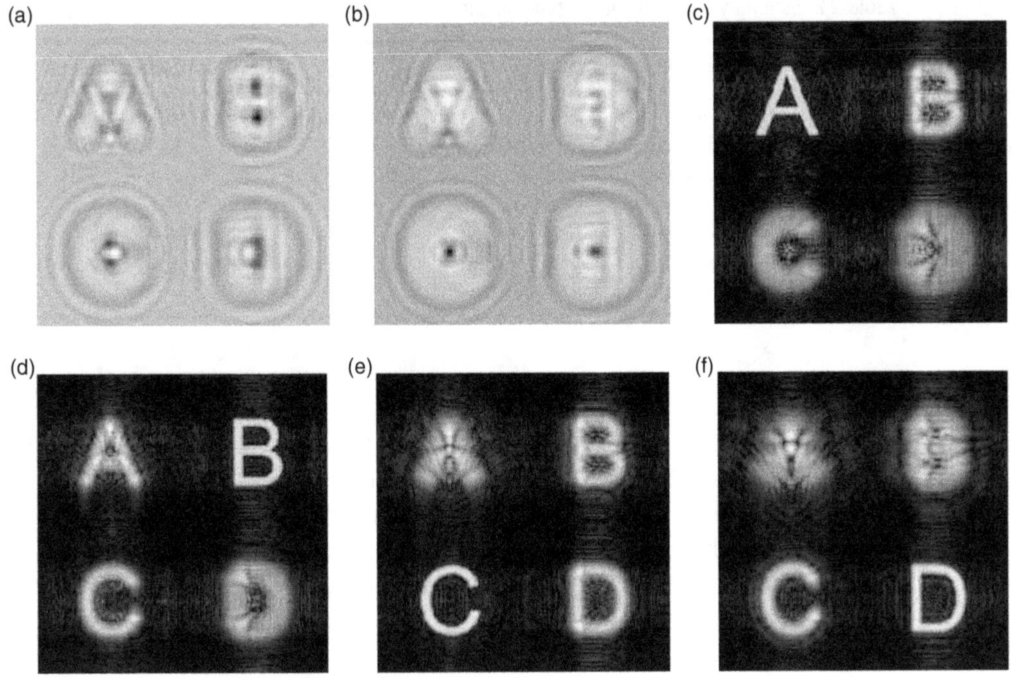

Figure 2.14 (a, b) Real and imaginary component of the warped WRP hologram of the prewarped image. (c–f) Reconstructed images of the hologram generated with the WWRP method at 0.020 m, 0.025 m, 0.030 m, and 0.035 m, respectively.

0.015 m from the image. The real and the imaginary components of the WRP hologram are shown in Figures 2.13(d) and 2.13(e), respectively. The reconstructed image is similar to the source image, which is not shown here.

Next, the WRP hologram is divided into four regions in accordance with the partitioning, as shown in Figure 2.13(b). In each region, the hologram fringes are warped with a factor $a_j|_{0\leq j\leq 3}$ following Eq. (2.29). The real and the imaginary parts of the warped holograms are shown in Figures 2.14(a) and 2.14(b). As shown in the figures, the fringe patterns for the four characters have been rescaled according to their distances between the source image and the hologram. The reconstructed images at the four focused planes $[0.020\text{ m}, 0.025\text{ m}, 0.030\text{ m}, 0.035\text{ m}]$ are shown in Figure 2.14(c–f). It can be seen that all four characters have been reconstructed to their original size.

2.11 MATLAB Simulation

This section includes the MATLAB source codes that are employed to conduct the simulations of the methods reported in this chapter. A summary of the simulation programs (each identified by the name of its main function call) and their correspondence to previous sections are listed in Table 2.2.

Table 2.2 Summary of the simulation programs in this chapter

Function name	Description of simulation	Associated sections
WRP	Computer-generated holograms with the wavefront diffraction plane (WRP) method	2.8
DS_WRP	Computer-generated holograms with the downsampled wavefront diffraction plane (DS-WRP) method	2.8
WWRP	Generating a hologram with the warped wavefront diffraction plane (WWRP) method	2.10

In the simulation on the WRP and the DS-WRP methods, the source image is the double-depth image HK1.jpg, with the intensity and the depth map shown in Figure 1.12(a,b). The image is evenly divided into a left half-plane and a right half-plane containing the characters H and K, respectively. The left half-plane is located at a distance of 0.07 m from the hologram plane, while the right half-plane is placed farther away at 0.08 m.

Simulation of the WWRP method is based on the source image "ABCD," which comprises four characters each located at a different axial distance from the hologram.

2.11.1 Simulation of Computer-Generated Hologram with the WRP Method

The object space is a double-depth image comprising a left section with the character H and a right section with the character K, which are located at 0.07 m and 0.08 m from the hologram plane, respectively. The WRP is parallel to, and at a distance of 0.06 m from, the hologram. A square support of 64 × 64 pixels is adopted in the computation. The MATLAB program, as shown in Table 2.3, has a main program "WRP" that loads the intensity of the double-depth image. The size of the WRP and the hologram are both set to twice that of the image. The function "Image_to_WRP" is called to compute the object waves on the WRP. Next, a Fresnel hologram is generated from the WRP with the function "WRP_to_hologram," from which the pair of images are reconstructed on their focused planes, displayed, and saved.

2.11.2 Simulation of Computer-Generated Hologram with the Downsampled WRP Method

The source code in Table 2.4 is used to generate a hologram with the DS-WRP method. The object space is a double-depth image comprising a left section with the character H and a right section with the character K, which are located at 0.07 m and 0.08 m from the hologram plane, respectively. The WRP is parallel to, and at a distance of 0.06 m from, the hologram. A square support of 64 × 64 pixels is adopted in the computation. The size of the WRP and the hologram are both set to twice the size of the image.

Table 2.3 Generating a hologram with the WRP method

```
function WRP()
    close all
    clc
    warning off

    fname='HK1';
    I=imread(strcat(fname,'.jpg'));    %Filename of intensity image
    I=im2double(I);
    [n m]=size(I(:,:));                %Size of source image

    psize=6.4e-6;                      %Pixel size of hologram=8.1 µm
    lamda=520e-9;                      %Wavelength=633 nm

    %Generate object wave of image on WRP
    %Square support [height,width]=[SN,SM]
    SN=64;
    SM=64;
    z=0.06;                            %Distance of WRP to hologram
    z1=0.01;                           %Distance of left half-plane of image to
                                         %WRP
    z2=0.02;                           %Distance of right half-plane of image
                                         %to WRP

    HWRP=Image_to_WRP(I,z1,z2,SN,SM,psize,lamda);

    figure,imshow(mat2gray(I));
    title('Source image');
    figure,imshow(mat2gray(real(HWRP)));
    title('Real part of WRP');
    figure,imshow(mat2gray(imag(HWRP)));
    title('Imag part of WRP');

    imwrite(mat2gray(real(HWRP)),'WRP_Real.jpg');
    imwrite(mat2gray(imag(HWRP)),'WRP_Imag.jpg');

    [N M]=size(HWRP(:,:));
    H=WRP_to_hologram(HWRP,z,psize,lamda);

    figure
    imshow(mat2gray(real(H)));         %Display real part of hologram
    title('Real part of the digital Fresnel hologram');
    figure,imshow(mat2gray(imag(H))); %Display imaginary part of hologram
    title('Imaginary part of the digital Fresnel hologram');

    imwrite(mat2gray(real(H)),'Holo_WRP_Real.jpg');
    imwrite(mat2gray(imag(H)),'Holo_WRP_Imag.jpg');

    %Compute reconstructed image at the two depth planes
    RES1=mat2gray(abs(Holo_recon(H,z+z1,psize,lamda)));
    RES2=mat2gray(abs(Holo_recon(H,z+z2,psize,lamda)));

    %Corp reconstructed image to size of source image
    sx=(M-m)/2;ex=(M+m)/2-1;
    sy=(N-n)/2;ey=(N+n)/2-1;
```

Table 2.3 (cont.)

```
    RES_Corp1=RES1(sy:ey,sx:ex);
    RES_Corp2=RES2(sy:ey,sx:ex);

    %Save reconstructed images at the two depth planes
    imwrite(RES_Corp1,'RES_WRP_D1.jpg');
    imwrite(RES_Corp2,'RES_WRP_D2.jpg');

    figure,imshow(RES_Corp1);           %Display reconstructed image at z0
    title('reconstructed image of the digital Fresnel hologram at z=z1');

    figure,imshow(RES_Corp2);           %Display reconstructed image at z0
    title('reconstructed image of the digital Fresnel hologram at z=z2');
end

function H=Image_to_WRP(I,z1,z2,SN,SM,psize,lamda)
    %Get size of input image
    [N M]=size(I(:,:));
    %===========================================================

    wn=2*pi/lamda;                      %Wave number

    %Declare 2-D array for storing the hologram
    %Size of hologram = Two times size of image
    H=zeros(2*N,2*M);

    %Declare meshgrid for calculating the object wave of a point source
    %Bounded by the square support (sx,sy) to (ex,ey)
    sx=-SM/2;
    sy=-SN/2;
    ex=SM/2-1;
    ey=SN/2-1;
    [kx ky]=meshgrid(sx:ex,sy:ey);
    kx=kx*psize;                        %Physical horizontal position
    ky=ky*psize;                        %Physical vertical position

    %Compute and accumulate the object wave of each object point
    wb=waitbar(0,'Please wait ...');
    for row=1:N
      y=row+N/2;
      for col=1:M
      x=col+M/2;
      A=I(row,col);                     %Read intensity of a point at (row,col),
                                          %and
                                        %shift to center of the hologram plane
     if A~=0                            %If intensity>0, it is an object point
        if col<=M/2                     %If left half-plane, depth = z1
           zd=z1;
        else
           zd=z2;                       %If right half-plane depth = z2
        end
        zsq=zd*zd;                      %Square of the depth value

        %Distance of object point to the WRP within area of coverage
```

Table 2.3 (cont.)

```
        r=sqrt(kx.^2+ky.^2+zsq);

        %Add the object waves of the object point to the WRP
        H(y+sy:y+ey,x+sx:x+ex)=H(y+sy:y+ey,x+sx:x+ex)+A*exp(j*wn*r);
     end

    end
    waitbar(row/N);
   end
   delete(wb);
end

function H=WRP_to_hologram(HWRP,z,psize,lamda)
   %size of WRP and hologram plane
   [N M]=size(HWRP(:,:));

   wn=2*pi/lamda;                         %Wave number

%Declare 2-D array for storing the hologram
H=zeros(N,M);

%Generate Fresnel zone plate
zsq=z*z;                                  %Square of the depth value z
[kx ky]=meshgrid(-M/2:M/2-1,-N/2:N/2-1);
kx=kx*psize;                              %Physical horizontal position
ky=ky*psize;                              %Physical vertical position
fzp=exp(1i*wn*sqrt(kx.^2+ky.^2+zsq));

A=fft2(HWRP);                             %Fourier transform of the hologram
B=fft2(fzp);                              %Fourier transform of the free space
                                            %impulse response

   H=fftshift(ifft2(A.*B));               %Compute hologram from WRP image
end

function IR=Holo_recon(H,z,psize,lamda)
    [N M]=size(H(:,:));                   %Size of hologram
    wn=2*pi/lamda;                        %Wave number

    %Generate Fresnel zone plate
    zsq=z*z;                              %Square of the depth value z
    [kx ky]=meshgrid(-M/2:M/2-1,-N/2:N/2-1);
    kx=kx*psize;                          %Physical horizontal position
    ky=ky*psize;                          %Physical vertical position
    r=sqrt(kx.^2+ky.^2+zsq);
    fzp_c=exp(-1i*wn*r);                  %Conjugate of FZP

    A=fft2(H);                            %Fourier transform of the hologram
    B=fft2(fzp_c);                        %Fourier transform of the free space
                                            %impulse response

    IR=fftshift(ifft2(A.*B));             %Compute reconstructed image
end
```

Table 2.4 Generating a hologram with the DS-WRP method

```
function DS_WRP()
    close all
    clc
    warning off

    fname='HK1';
    I=imread(strcat(fname,'.jpg'));  %Filename of intensity image
    I=im2double(I);
    [n m]=size(I(:,:));               %Size of source image

    psize=6.4e-6;                     %Pixel size of hologram=8.1 µm
    lamda=520e-9;                     %Wavelength=633 nm

   %Generate object wave of image on WRP
    %Square support [height,width]=[SN,SM]
    SN=64;
    SM=64;
    z=0.06;                           %Distance of WRP to hologram
    z1=0.01;                          %Distance of left half-plane of image to
                                        %WRP
    z2=0.02;                          %Distance of right half-plane of image
                                        %to WRP

    DS=input('Enter downsampling factor ');

    HWRP=Image_to_WRP(I,z1,z2,DS,SN,SM,psize,lamda);

    figure,imshow(mat2gray(I));
    title('Source image');
    figure,imshow(mat2gray(real(HWRP)));
    title('Real part of WRP');
    figure,imshow(mat2gray(imag(HWRP)));
    title('Imag part of WRP');

    imwrite(mat2gray(real(HWRP)),'WRP_DS_Real.jpg');
    imwrite(mat2gray(imag(HWRP)),'WRP_DS_Imag.jpg');

    [N M]=size(HWRP(:,:));

    H=WRP_to_hologram(HWRP,z,psize,lamda);

    figure,imshow(mat2gray(real(H))); %Display real part of hologram
    title('Real part of the digital Fresnel hologram');
    figure,imshow(mat2gray(imag(H))); %Display imaginary part of hologram
    title('Imaginary part of the digital Fresnel hologram');

    imwrite(mat2gray(real(H)),'Holo_WRP_DS_Real.jpg');
    imwrite(mat2gray(imag(H)),'Holo_WRP_DS_Imag.jpg');

    %Compute reconstructed image at the two depth planes
    RES1=mat2gray(abs(Holo_recon(H,z+z1,psize,lamda)));
    RES2=mat2gray(abs(Holo_recon(H,z+z2,psize,lamda)));

    %Corp reconstructed image to size of source image
    sx=(M-m)/2;ex=(M+m)/2-1;
    sy=(N-n)/2;ey=(N+n)/2-1;
```

Table 2.4 (cont.)

```
    RES_Corp1=RES1(sy:ey,sx:ex);
    RES_Corp2=RES2(sy:ey,sx:ex);

    %Save Reconstructed images at the two depth planes
    imwrite(RES_Corp1,'RES_WRP_DS_D1.jpg');
    imwrite(RES_Corp2,'RES_WRP_DS_D2.jpg');

    figure,imshow(RES_Corp1);          %Display reconstructed image at z0
    title('reconstructed image of the digital Fresnel hologram at z=z1');

    figure,imshow(RES_Corp2);          %Display reconstructed image at z0
    title('reconstructed image of the digital Fresnel hologram at z=z2');

end

function H=Image_to_WRP(I,z1,z2,DS,SN,SM,psize,lamda)
  %Get size of input image
  [N M]=size(I(:,:));
  %=======================================================

  wn=2*pi/lamda;                       %Wave number

  %Declare 2-D array for storing the hologram
  %Size of hologram = Two times size of image
  H=zeros(2*N,2*M);

  %Declare meshgrid for calculating the object wave of a point source
  %Bounded by the square support (sx,sy) to (ex,ey)
  sx=-SM/2;
  sy=-SN/2;
  ex=SM/2-1;
  ey=SN/2-1;
  [kx ky]=meshgrid(sx:ex,sy:ey);
  kx=kx*psize;                         %Physical horizontal position
  ky=ky*psize;                         %Physical vertical position

  %Compute and accumulate the object wave of each object point
  wb=waitbar(0,'Please wait...');
  for row=1:N
    y=row+N/2;
    p=mod(row,DS);
    for col=1:M
        x=col+M/2;
        q=mod(col,DS);
        A=I(row,col);                  %Read intensity of a point at (row,col)
                                       %Shift to center of hologram plane

   %If the intensity of the sampled point>0, it is an object point
      if A~=0 && p==0 && q==0
        if col<=M/2                    %Left half-plane
          zd=z1;                       %Depth of left half-plane
        else
            zd=z2;                     %Depth of right half-plane
        end
        zsq=zd*zd;                     %Square of the depth value
```

Table 2.4 (cont.)

```
          %Distance of object point to the WRP within area of coverage
          r=sqrt(kx.^2+ky.^2+zsq);
          %Add the object waves of the object point to the WRP
          H(y+sy:y+ey,x+sx:x+ex)=H(y+sy:y+ey,x+sx:x+ex)+A*exp(j*wn*r);
       end
       end
       waitbar(row/N);
    end
   delete(wb);
end

function H=WRP_to_hologram(HWRP,z,psize,lamda)

%size of WRP and hologram plane
[N M]=size(HWRP(:,:));

wn=2*pi/lamda;                              %Wave number

%Declare 2-D array for storing the
 hologram

H=zeros(N,M);

%Generate Fresnel zone plate
zsq=z*z;                                    %Square of the depth value z
[kx ky]=meshgrid(-M/2:M/2-1,-N/2:N/2-1);
kx=kx*psize;                                %Physical horizontal position
ky=ky*psize;                                %Physical vertical position
fzp=exp(1i*wn*sqrt(kx.^2+ky.^2+zsq));

A=fft2(HWRP);                               %Fourier transform of the hologram
B=fft2(fzp);                                %Fourier transform of the free space
                                              %impulse response
H=fftshift(ifft2(A.*B));                    %Compute hologram from the WRP image

end

function IR=Holo_recon(H,z,psize,lamda)

[N M]=size(H(:,:));                         %Size of hologram

wn=2*pi/lamda;                              %Wave number

%Generate Fresnel Zone Plate
zsq=z*z;                                    %Square of the depth value z
[kx ky]=meshgrid(-M/2:M/2-1,-N/2:N/2-1);
kx=kx*psize;                                %Physical horizontal position
ky=ky*psize;                                %Physical vertical position
r=sqrt(kx.^2+ky.^2+zsq);
fzp_c=exp(-1i*wn*r);                        %Conjugate of FZP

A=fft2(H);                                  %Fourier transform of the hologram
B=fft2(fzp_c);                              %Fourier transform of the free space
                                              %impulse response
IR=fftshift(ifft2(A.*B));                   %Compute reconstructed image
end
```

First the main program "DS_WRP" loads the intensity and depth map of the double-depth image. After the image is loaded, the user is prompted to enter a downsampling factor, DS. The function "Image_to_WRP" is called, with which the intensity image is downsampled, and a WRP is generated from the downsampled image and its depth map. Next, a Fresnel hologram is generated from the WRP with the function "WRP_to_hologram," from which the pair of images are reconstructed on their focused planes.

2.11.3 Simulation of Computer-Generated Hologram with the WWRP Method

The source code in Table 2.5 is used to generate the hologram of a multiple-depth object. The object space is a square image that is evenly partitioned into four sections, each occupying a corner of the image, and containing one of the four characters {A,B, C,D}. The four sections are located at 0.020 m, 0.025 m, 0.030 m, and 0.035 m from the WRP, which is parallel to, and at a distance of 0.06 m from, the hologram.

The distance between the prewarp image and the WRP is 0.015 m, which is nearer to the WRP than the actual distances of each section. As such, all the sections on the source image, as well as its hologram on the WRP, have to be prewarped and enlarged, respectively.

In the MATLAB simulation, the function "WWRP" loads the intensity of the source image, and the function "Prewarp_Image" is called to prewarp each section of the image according to its distance from the WRP. Next, the WRP is generated from the prewarped image. The size of the WRP is set to be identical to that of the image. Subsequently, the function "Warp_Hologram" is executed to enlarge the fringe pattern of each section of the WRP hologram, according to the distance between the corresponding section and the WRP. Finally, each section of the image is reconstructed at its focused distance from the WRP hologram, displayed, and saved.

Note that for the sake of simplicity, the simulation does not include the forward and backward conversion of the WRP to and from the hologram.

2.12 Summary

In this chapter, a number of fast methods for generating digital Fresnel holograms were introduced. For an object comprising a small number of depth planes, the layer-based method that is implemented in the Fourier space is preferred for fast generation of its hologram. The point-based method, on the other hand, is more suitable for generating objects with a large number of object points that are scattering over a wide range of distances from the hologram plane. To increase the speed of hologram generation with the point-based method, different variants and sizes of LUT algorithms to trade-off computation time are described.

In the second part of the chapter, a number of methods based on the concept of the wavefront recording plane (WPR) are presented. Being different from the LUT approach, the WRP methods speed up the hologram-generation process through simplifying the hologram-formation process.

Table 2.5 Generating the hologram of a multiple-depth object with the WWRP method

```
function WWRP()
  close all
  %Filename of intensity image
  fname='ABCD.jpg';
  %================= Read intensity and depth images =====================
  %Read intensity image and store in a 2-D array of type 'double'
    IMG=imread(fname);
    imshow(IMG);
    IMG=im2double(IMG);

    [N M]=size(IMG(:,:));
   %==============================================================

   %Optical settings for generating hologram
   psize=6.4e-6;
   lamda=520e-9;
   z0=0.015;                              %Distance of WRP from image plane
   H_DIM=N;                               %Size of hologram = size of image

  %Partition the source image into four sections, each at certain distance from
  %the image plane

   %Prewarp the section of the upper left corner
   sx=1;sy=1;ex=N/2;ey=M/2;
   z1=0.02;scale=sqrt(z0/z1);
   IMG=Prewarp_Image(IMG,sx,sy,ex,ey,scale);

   %Prewarp the section of the upper right corner
   sx=N/2+1;sy=1;ex=N;ey=M/2;
   z2=0.025;scale=sqrt(z0/z2);
   IMG=Prewarp_Image(IMG,sx,sy,ex,ey,scale);

   %Prewarp the section of the lower left corner
   sx=1;sy=M/2+1;ex=N/2;ey=M;
   z3=0.03;scale=sqrt(z0/z3);
   IMG=Prewarp_Image(IMG,sx,sy,ex,ey,scale);

   %Prewarp the section of the lower right corner
   sx=N/2+1;sy=M/2+1;ex=N;ey=M;
   z4=0.035;scale=sqrt(z0/z4);
   IMG=Prewarp_Image(IMG,sx,sy,ex,ey,scale);

   %Display and save the prewarped image
   figure;imshow(IMG);title('Warped image');
   imwrite(IMG,'Warped_image.jpg');

   %Compute WRP hologram of source image
   H=CGH_gen(IMG,z0,psize,lamda,H_DIM);

   %Display and save WRP hologram
   HR=mat2gray(real(H));
   HI=mat2gray(imag(H));
   figure; imshow(HR);title('Real part of hologram');
```

Table 2.5 (cont.)

```
    imwrite(HR,'Holo_WRP_Real.jpg');
    figure; imshow(HI);title('Imaginary part of hologram');
    imwrite(HI,'Holo_WRP_Imag.jpg');

    %Enlarge the WRP hologram of each section, according to its depth
    %Magnify the section of the WRP hologram of the upper left corner
    sx=1;sy=1;ex=H_DIM/2;ey=H_DIM/2;
    z1=0.02;scale=sqrt(z0/z1);
    H=Warp_Hologram(H,sx,sy,ex,ey,scale);

    %Magnify the section of the WRP hologram of the upper right corner
    sx=H_DIM/2+1;sy=1;ex=H_DIM;ey=H_DIM/2;
    z2=0.025;scale=sqrt(z0/z2);
    H=Warp_Hologram(H,sx,sy,ex,ey,scale);

    %Magnify the section of the WRP hologram of the lower left corner
    sx=1;sy=N/2+1;ex=M/2;ey=N;
    z3=0.03;scale=sqrt(z0/z3);
    H=Warp_Hologram(H,sx,sy,ex,ey,scale);

    %Magnify the section of the WRP hologram of the lower right corner
    sx=N/2+1;sy=M/2+1;ex=N;ey=M;
    z4=0.035;scale=sqrt(z0/z4);
    H=Warp_Hologram(H,sx,sy,ex,ey,scale);

    %Display and save real and imaginary components of WRP hologram
    HRW=mat2gray(real(H));
    HIW=mat2gray(imag(H));
    figure; imshow(HRW);title('Real part of warped hologram');
    imwrite(HRW,'Holo_WWRP_Real.jpg');
    figure; imshow(HIW);title('Imaginary part of warped hologram');
    imwrite(HIW,'Holo_WWRP_Imag.jpg');

    %Reconstruct and save each section from the WRP hologram
    RES1=mat2gray(abs(CGH_recon(H,z1,psize,lamda)));
    figure;imshow(RES1);title('Reconstructed image at 0.020 m');
    imwrite(RES1,strcat('RES_WWRP_D1.jpg'));
    RES2=mat2gray(abs(CGH_recon(H,z2,psize,lamda)));
    figure;imshow(RES2);title('Reconstructed image at 0.025 m');
    imwrite(RES2,strcat('RES_WWRP_D2.jpg'));
    RES3=mat2gray(abs(CGH_recon(H,z3,psize,lamda)));
    figure;imshow(RES3);title('Reconstructed image at 0.030 m');
    imwrite(RES3,strcat('RES_WWRP_D3.jpg'));
    RES4=mat2gray(abs(CGH_recon(H,z4,psize,lamda)));
    figure;imshow(RES4);title('Reconstructed image at 0.035 m');
    imwrite(RES4,strcat('RES_WWRP_D4.jpg'));

end

function HU=Warp_Hologram(H,sx,sy,ex,ey,scale)
  %Region in hologram bounded by {sx,sy,ex,ey}
  HU=H;
  HU(sy:ey,sx:ex)=0;
```

Table 2.5 (cont.)

```
    %Centroid of region on the WRP
    cx=(ex+sx)/2;
    cy=(ey+sy)/2;

    for row=sy:ey
      for col=sx:ex
         y=(row-cy);
         x=(col-cx);
         y1=round(y*scale+cy);          %Scale the specified WRP region
                                          %vertically
         x1=round(x*scale+cx);          %Scale the specified WRP region
                                          %horizontally

         HU(row,col)=H(y1,x1);          %Replace the specified WRP region with
                                          %the scaled version
      end
    end
end

function IR=Prewarp_Image(IMG,sx,sy,ex,ey,scale)
   %Region in image bounded by {sx,sy,ex,ey}
   IR=IMG;
   IR(sy:ey,sx:ex)=0;
   SC=scale;

   %Centroid of region on the image
   cx=(ex+sx)/2
   cy=(ey+sy)/2
     for row=sy:ey
      for col=sx:ex
      y=(row-cy);
      x=(col-cx);
      y1=round(y/scale+cy);             %Scale the specified image region
                                          %vertically
      x1=round(x/scale+cx);             %Scale the specified image region
                                          %horizontally
      if x1>sx && x1<ex && y1>sy &&y1<ey
        IR(row,col)=IMG(y1,x1);
      end
     end
   end
end

function H=CGH_gen(IMG,z,psize,lamda,H_DIM)
[M N]=size(IMG(:,:));
wn=2*pi/lamda;                          %Wave number
%Declare 2-D array for storing the hologram
H=zeros(H_DIM,H_DIM);

%Generate Fresnel Zone Plate
z2=z*z;                                 %Square of the depth value z
[kx ky]=meshgrid(-H_DIM/2:H_DIM/2-1,-H_DIM/2:H_DIM/2-1);
kx=kx*psize;                            %Physical horizontal position
```

Table 2.5 (cont.)

```
ky=ky*psize;                                %Physical vertical position
fzp=exp(1i*wn*(kx.^2+ky.^2)/2/z);

%Generate the WRP hologram of the image
A=fft2(IMG);                                %Fourier transform of the hologram
B=fft2(fzp);                                %Fourier transform of the free space
                                              %impulse response
H=fftshift(ifft2(A.*B));

end

function IR=CGH_recon(H,z,psize,lamda)

[N M]=size(H(:,:));                         %Size of hologram

wn=2*pi/lamda;                              %Wave number

%Generate Fresnel Zone Plate
z2=z*z;                                     %Square of the depth value z
[kx ky]=meshgrid(-M/2:M/2-1,-N/2:N/2-1);
kx=kx*psize;                                %Physical horizontal position
ky=ky*psize;                                %Physical vertical position
fzp=exp(-1i*wn*(kx.^2+ky.^2)/2/z);

A=fft2(H);                                  %Fourier transform of the hologram
B=fft2(fzp);                                %Fourier transform of the free space
                                              %impulse response
IR=fftshift(ifft2(A.*B));                   %Compute reconstructed image
end
```

Instead of generating the full hologram for each object point, only a small area of fringe patterns is computed on a WRP that is at close proximity to the object space. The computation time, therefore, is substantially reduced. Further enhancement on the computation speed is attained with the warped wavefront recording plane (WWRP) method. A 3-D object image is decomposed into a 2-D intensity image and a depth map. The intensity image is used to generate an interim hologram on a WRP. Different regions of the hologram fringes on the WRP are subsequently resized according to their distances (obtained from the depth map) from the hologram plane to generate a hologram from the WRP.

Exercises

E.2.1 Given $\lambda = 0.52$ μm, $\delta_d = 6.4$ μm, and H_DIM $= 2048$, perform a simulation with MATLAB to compare the computation time between generating a hologram of the double-depth image in Figure 1.12 with the layer-based method, based on (a) convolution, and (b) fast Fourier transform. From the results, evaluate the pros and cons of these two methods in the generation of digital Fresnel holograms.

E.2.2 Perform a literature study on reference [2], and write a MATLAB program to generate a digital Fresnel hologram of the double-depth image in Figure 1.12 with the N-LUT method.

E.2.3 Based on the MATLAB programs in Sections 2.11.1 and 2.11.2 that apply the WRP method to generate the digital Fresnel hologram of the double-depth image in Figure 1.12, evaluate the effects of the following parameters on the computation time, and the quality of the reconstructed images of the hologram.
a. Side length τ of the square support.
b. Distances z_0 and z_1 between the pair of depth planes and the WRP.
c. Downsampling factor on the intensity of the double-depth image.

E.2.4 Perform a literature study on references [19–22], and extend the MATLAB program in Section 2.11.2 to generate a digital Fresnel hologram of an image with four depth planes with the multi-depth WRP method. The intensity of the image is shown in Figure 2.13(a), and the axial distances of characters "A," "B," "C," and "D" from the hologram are 0.06 m, 0.07 m, 0.25 m, and 0.26 m, respectively. The optical settings are given in E.2.1

References

[1] M. Lucente, "Interactive computation of holograms using a look-up table," *J. Electron. Imaging*, vol. 2, pp. 28–34 (1993).

[2] S. Kim and E. Kim, "Effective generation of digital holograms of three-dimensional objects using a novel look-up table method," *Appl. Opt.*, vol. 47, pp. D55–D62 (2008).

[3] S. Kim, J. Kim, and E. Kim, "Effective reduction of the novel look-up table memory size based on a relationship between the pixel pitch and reconstruction distance of a computer-generated hologram," *Appl. Opt.*, vol. 50, pp. 3375–3382 (2011).

[4] Z. Yang, Q. Fan, Y. Zhang, J. Liu, and J. Zhou, "A new method for producing computer generated holograms," *J. Opt.*, vol. 14, 095702 (2012).

[5] J. Bresenham, "A linear algorithm for incremental digital display of circular arcs," *Commun. ACM*, vol. 20, pp. 100–106 (1977).

[6] T. Nishitsuji, T. Shimobaba, T. Kakue, N. Masuda, and T. Ito, "Fast calculation of computer-generated hologram using the circular symmetry of zone plates," *Opt. Express*, vol. 20, pp. 27496–27502 (2012).

[7] E. Andres, "Discrete circles, rings and spheres," *Comput. Graphics*, vol. 18, pp. 695–706 (1994).

[8] Y. Pan, X. Xu, S. Solanki, X. Liang, R. Tanjung, C. Tan, and T. Chong, "Fast CGH computation using S-LUT on GPU," *Opt. Express*, vol. 17, pp. 18543–18555 (2009).

[9] J. Jia, Y. Wang, J. Liu, X. Li, Y. Pan, Z. Sun, B. Zhang, Q. Zhao, and W. Jiang, "Reducing the memory usage for effective computer-generated hologram calculation using compressed look-up table in full-color holographic display," *Appl. Opt.*, vol. 52, pp. 1404–1412 (2013).

[10] C. Gao, J. Liu, X. Li, G. Xue, J. Jia, and Y. Wang, "Accurate compressed look up table method for CGH in 3D holographic display," *Opt. Express*, vol. 23, pp. 33194–33204 (2015).

[11] T. Yamaguchi and H. Yoshikawa, "Real time calculation for holographic video display," *Proc. SPIE*, vol. 6136, p. 6136T (2006).

[12] T. Shimobaba, N. Masuda, and T. Ito, "Simple and fast calculation algorithm for computer-generated hologram with wavefront recording plane," *Opt. Lett.*, vol. 34, pp. 3133–3135 (2009).

[13] T. Shimobaba, H. Nakayama, N. Masuda, and T. Ito, "Rapid calculation algorithm of Fresnel computer-generated-hologram using look-up table and wavefront-recording plane methods for three-dimensional display," *Opt. Express*, vol. 18, pp. 19504–19509 (2010).

[14] T. Nishitsuji, T. Shimobaba, T. Kakue, N. Masuda, and T. Ito, "Fast calculation of computer-generated hologram using the circular symmetry of zone plates," *Opt. Express*, vol. 20, pp. 27496–27502 (2012).

[15] T. Shimobaba and T. Ito, "Fast generation of computer-generated holograms using wavelet shrinkage," *Opt. Express*, vol. 25, pp. 77–87 (2017).

[16] D. Arai, T. Shimobaba, T. Nishitsuji, T. Kakue, N. Masuda, and T. Ito, "An accelerated hologram calculation using the wavefront recording plane method and wavelet transform," *Opt. Commun.*, vol. 393, pp. 107–112 (2017).

[17] N. Okada, T. Shimobaba, Y. Ichihashi, R. Oi, K. Yamamoto, T. Kakue, and T. Ito, "Fast calculation of a computer-generated hologram for RGB and depth images using a wavefront recording plane method," *Photon. Lett. Poland*, vol. 6, pp. 90–92 (2014).

[18] J. Weng, T. Shimobaba, N. Okada, H. Nakayama, M. Oikawa, N. Masuda, and T. Ito, "Generation of real-time large computer generated hologram using wavefront recording method," *Opt. Express*, vol. 20, pp. 4018–4023 (2012).

[19] A. Phan, M. Piao, S. Gil, and N. Kim, "Generation speed and reconstructed image quality enhancement of a long-depth object using double wavefront recording planes and a GPU," *Appl. Opt.*, vol. 53, pp. 4817–4824 (2014).

[20] A.-H. Phan, M.A. Alam, S.-H. Jeon, J.-H. Lee, and N. Kim "Fast hologram generation of long-depth object using multiple wavefront recording planes," Proc. SPIE 9006, Practical Holography XXVIII: Materials and Applications, 90061 (2014).

[21] A. Symeonidou, D. Blinder, A. Munteanu, and P. Schelkens, "Computer-generated holograms by multiple wavefront recording plane method with occlusion culling," *Opt. Express*, vol. 23, pp. 22149–22161 (2015).

[22] N. Hasegawa, T. Shimobaba, T. Kakue, and T. Ito, "Acceleration of hologram generation by optimizing the arrangement of wavefront recording planes," *Appl. Opt.*, vol. 56, pp. A97–A103 (2017).

[23] D. Arai, T. Shimobaba, K. Murano, Y. Endo, R. Hirayama, D. Hiyama, T. Kakue, and T. Ito, "Acceleration of computer-generated holograms using tilted wavefront recording plane method," *Opt. Express*, vol. 23, pp. 1740–1747 (2015).

[24] P.W.M. Tsang, W.K. Cheung, T.-C. Poon, and C. Zhou, "Holographic video at 40 frames per second for 4-million object points," *Opt. Express*, vol. 19, pp. 15205–15211 (2011).

[25] P.W.M. Tsang and T.-C. Poon, "Fast generation of digital holograms based on warping of the wavefront recording plane," *Opt. Express*, vol. 23, pp. 7667–7673 (2015).

3 Generation of Phase-Only Fresnel Hologram

3.1 General View on Holographic Display System

Some of the important methods that are employed in the generation of digital Fresnel holograms have been introduced in the previous chapters. As mentioned before, a digital hologram is merely an array of complex-valued data, which is not meaningful to the real world unless it can be converted into a physical hologram and used to display a visible holographic 3-D image. An early approach to forming a physical hologram from digital holographic data was to generate a real off-axis hologram (see Sections 1.4 and 1.7.1) and to print it with a high-resolution printer. Printer technology has advanced significantly in the past two decades; it is now possible to hardcopy holograms with resolution of over 2400 dots-per-inch (dpi) through a commodity laser printer. Even higher printing resolution can be achieved with professional image setters or holographic fringe printers. In general, the higher the resolution of the printing device, the better will be the quality of the reconstructed image of a hologram. Despite these encouraging developments in hologram printing, the content of a printed hologram is not refreshable and can only record a static scene.

The advancement of electronic display technology has provided the green light to the display of dynamic holographic images. At the time of writing, display devices such as liquid-crystal-on-silicon (LCoS), with a pixel size of around 3 μm are available at an affordable price. Although the size of these devices is generally small, they are sufficient for showing a medium-size hologram with good visual quality. In the study of holography, high-resolution electronic displays are generally referred to as spatial light modulators (SLMs), meaning that the optical properties of light (such as polarization, phase, and amplitude) could be changed when the optical beam is transmitted or reflected from the device. However, SLMs developed to date are only capable of changing either the amplitude or the phase component of an optical signal. An SLM that is configured to modulate the amplitude or phase component is referred to as an amplitude-only SLM or a phase-only SLM, respectively. In either configuration, a single SLM is inadequate for displaying a hologram comprising complex-valued pixels.

3.1.1 Dual SLM Holographic Display System

To overcome this problem, a holographic display formed by the cascade of a pair of SLMs is proposed by C. Stolz et al. [1] and R. Tudela et al. [2]. The optical setup is

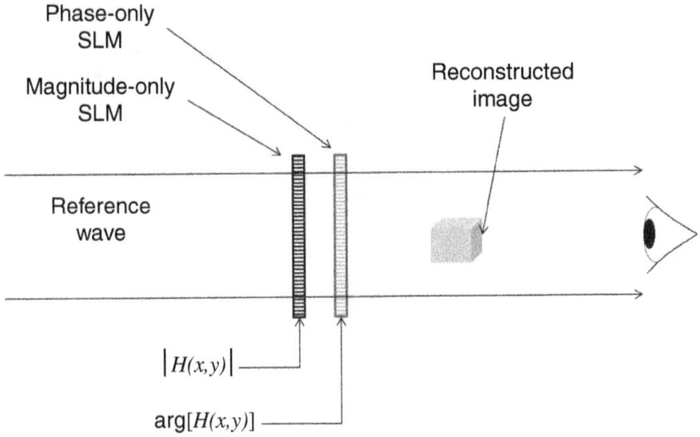

Figure 3.1 Concept of a dual, magnitude–phase coupled SLM holographic display system.

shown in Figure 3.1. The first SLM modulates the amplitude of the illumination beam according to the amplitude component of the hologram, while the second SLM modulates the phase of the optical wave according to the phase component of the hologram. The light wave that emerges from the two SLMs will be equivalent to the holographic signal emitted from the hologram, from which a reconstructed image of the 3-D scene can be observed. Although this is a straightforward implementation, the setup is bulky and expensive as two SLMs are involved. Due to the fine resolution of the display devices, precise optical alignment is required to ensure the correct amplitude and phase modulation of every light beam that passes through every pair of corresponding pixels of the two SLMs. Nevertheless, the method is effective and further research works have been conducted along this direction. An alternative two-SLM approach has been reported by L. Zhu and J. Wang [3] and A. Siemion et al. [4], who decomposed a complex-valued hologram into a pair of phase components, and displayed them with two cascaded phase-only SLMs. The concept has been extended by M. Makowski and his co-workers for displaying color Fourier holograms [5].

3.1.2 Split SLM Holographic Display System

The optical setup in Figure 3.1 can be much simplified if only a single SLM is used in the display system. Such an attempt has been made by Liu et al. [6] with the setup shown in Figure 3.2. Instead of separating the real and the imaginary components into two different SLMs, they are multiplexed on a single SLM. Referring to Figure 3.2, an amplitude-only SLM is uniformly split into two areas. The real component $H_R(m,n)$ and the imaginary component $H_I(m,n)$ of a complex-valued hologram are each displayed on one of these two areas. When the SLM is illuminated with a coherent wave, the optical beam that passes through each section is amplitude-modulated by the hologram pixels on it. The modulated light waves from the two sections are then merged to form the holographic signal of the hologram through some optical devices.

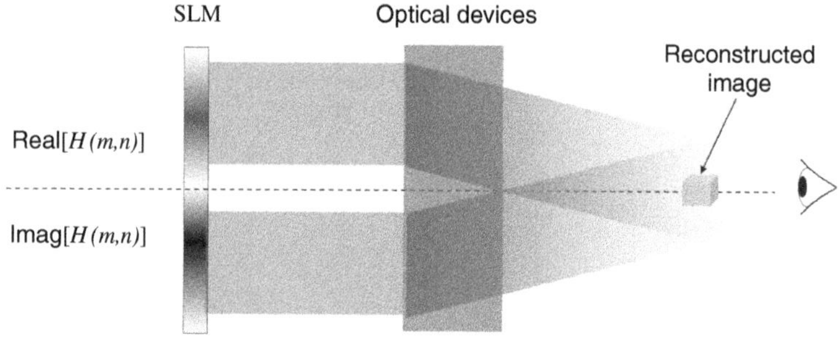

Figure 3.2 A partitioned, single SLM holographic display system.

From the holographic signal, a visible 3-D image of the object scene that is represented by the original complex hologram will be reconstructed.

A different solution, which is also based on the split SLM configuration, has been proposed by H. Song et al. [7]. A hologram is decomposed into a pair of phase components, with each of them being displayed in separate sections on a phase-only SLM. Similar to [6], the SLM is illuminated with a coherent beam, and the optical waves emitted from the two sections are merged optically to give the complex-valued holographic signal.

There are two major problems associated with the split SLM display system. First, as a single SLM is used to display the two components of a complex-valued hologram, the display area is reduced by two times. The size of the hologram, and hence the space–bandwidth product, are decreased by the same factor. As a result, the field of vision of a hologram displayed on a split SLM is smaller than that of a single SLM. Second, optical accessories such as lenses and high-resolution gratings are required to combine the optical wave from the two sections of the SLM. The alignment of the SLM and the optical accessories has to be very precise. Otherwise, the two components of the hologram cannot be integrated to recover the original holographic signal and are heavily distorted on the reconstructed image. In view of these issues, it is desirable if a hologram can be encapsulated into a single component (either the amplitude or the phase component), so that it can be displayed with a single, non-partitioned SLM.

3.1.3 Amplitude-Only SLM Holographic Display System

In the study of optical holography in Section 1.4, a hologram can be represented solely by its magnitude component if it is mixed with an inclined reference wave. This is the classic way of recording hologram onto a photographic film. Following the same principle, an amplitude-only SLM can be used to display an off-axis hologram as the values of all the pixels are real quantities. As explained in Section 1.4, there are two major drawbacks. First, the reconstructed image of an off-axis hologram is accompanied by the zero-order beam and the twin image, both of which may contaminate the virtual image if the pixel size of the SLM is not fine enough. Second, the optical efficiency of an amplitude-only hologram is rather low, as hologram pixels are translucent (i.e., having transmissivity between opaque and transparent), which blocks or fails to reflect some of the light energy.

A second solution is to display the phase component of a hologram alone by modulating the phase angle of the optical beam that impinges on it. Being different from an amplitude-only hologram, the optical efficiency of a phase-only hologram is high, as the transmissivity of hologram pixels is close to 100%, and only the phase angle of the optical beam is changed. However, direct removal of the magnitude component of a hologram will lead to heavy distortion in the reconstructed image. To overcome this problem, the hologram has to be generated in a way that the magnitude component is more or less homogeneous. Such kind of holograms are referred to as phase-only holograms. In the past few decades there have been numerous research efforts to explore methods for generating phase-only holograms. In the subsequent part of this chapter, different methods that have been developed in recent years on the generation of phase-only Fresnel hologram will be reviewed.

3.2 Iterative Method for Generating Phase-Only Holograms

3.2.1 Generating Phase-Only Hologram for a Single-Depth Image

The Gerchberg–Saxton algorithm, often referred to as GSA [8], is by far one of the most popular approaches for generating a digital phase-only hologram. The algorithm, named after its two inventors, Gerchberg and Saxton, was originally used to retrieve the phase of a pair of related light distributions. The principles of GSA have been extended to the generation of digital phase-only Fourier holograms. A Fourier hologram is a complex-valued image that is obtained from the Fourier transform of an image. Similar to what was mentioned in Section 3.1, it will be desirable to convert it into a phase-only component to increase the optical efficiency.

Figure 3.3 depicts a variation of the GSA known as the "iterative Fourier transform algorithm" (IFTA), which is applied to generate a phase-only Fourier hologram. The pair of light distributions are the intensity $I(m,n)$ of the source object on the image plane and its digital Fourier hologram $H(m,n)$ on the hologram plane. The objective is to derive a hologram so that the reconstructed image of its phase component $H_P(m,n)$ is similar to $I(m,n)$. Intuitively, this seems to contradict the theory of holography as the phase component of a hologram alone is insufficient to recover the original image. However, this dilemma can be alleviated if $H_P(m,n)$ is being solely required to reconstruct a modified version of $I(m,n)$ given by $J(m,n) \approx I(m,n)e^{i\phi(m,n)}$, where $e^{i\phi(m,n)}$ is a pure phase term. The latter provides an extra dimension of freedom for partially absorbing the distortion caused by dropping the magnitude component of the hologram. As human perception only responds to the intensity of light, introduction of the extra phase term does not affect the appearance of the image. In IFTA, computation of a phase-only hologram $H_P(m,n)$ is obtained through repetitive rounds of iterations, explained below.

To start with, let t denote the current round of iteration, and $G_t(m,n) = e^{i\phi_t(m,n)}$ is the phase-only hologram obtained in the current epoch. Initially, at $t = 0$, each pixel in the phase-only hologram $\phi_{t=0}(m,n)$ is assigned a random value within the range $[-\pi, \pi]$.

Next, the phase-only hologram is inverse Fourier transformed to a complex image $J_t(m,n) = |J_t(m,n)|e^{i\theta_t(m,n)}$ on the image plane, as given by

$$J_t(m,n) = |J_t(m,n)|e^{i\theta_t(m,n)} = \mathcal{F}^{-1}[G_t(m,n)]. \tag{3.1}$$

Subsequently, an amplitude constraint is imposed with which the magnitude of $J_t(m,n)$ is replaced by $I(m,n)$, resulting in a phase-added image $IP_t(m,n) = I(m,n)e^{i\phi_t(m,n)}$. The phase-added image $IP_t(m,n)$ is then Fourier transformed into a Fourier hologram $H_t(m,n) = |H_t(m,n)|e^{i\phi_{H;t}(m,n)}$:

$$H_t(m,n) = |H_t(m,n)|e^{i\phi_{H;t}(m,n)} = \mathcal{F}[IP_t(m,n)], \tag{3.2}$$

and the phase component $\phi_{H;t}(m,n)$ of the hologram is retained as the phase-only hologram $G_{t+1}(m,n)$, completing the current round of iteration. The conversion of the signals from the image plane to the hologram plane, and vice versa, is commonly referred to as forward and backward propagation, respectively.

The above process is repeated for a certain number of epochs, or until the difference between $I(m,n)$ and $|J_t(m,n)|$ is below a preset threshold. Note that in Eq. (3.2), the amplitude constraint is imposed so that the magnitude of the image to be converted to the hologram is always enforced to be identical to the original image $I(m,n)$. The phase term $e^{i\phi_t(m,n)}$ is allowed to vary freely according to the recent round of iteration. This has the effect of absorbing part of the errors caused by retaining only the phase component of the hologram. Upon completion of the process at the Tth epoch, the phase component $G_T(m,n) = e^{i\phi_T(m,n)}$ is taken to be the phase-only hologram $H_P(m,n)$. With more iterations, the reconstructed image of $H_P(m,n)$ will become increasingly similar to the source image $I(m,n)$.

The IFTA for generating Fourier phase-only holograms can be easily extended to the generation of Fresnel phase-only holograms, simply by replacing the forward/inverse Fourier transform with the forward/inverse Fresnel transform in Figure 3.3. This IFTA approach reported in [9].

There are several major disadvantages of the IFTA. First, for an arbitrary source image it is almost impossible to estimate how many rounds of iterations are required to generate the desired phase-only hologram (i.e., one with a reconstructed image that is close to the original one). Second, the reconstructed image is quite noisy as a result of the distortions caused by removing the magnitude component of the hologram. Third, the errors that can be absorbed into the phase term $e^{i\phi_T(m,n)}$ are quite limited. As a result, it is unlikely that the quality of the reconstructed image of $H_P(m,n)$ can be improved significantly simply by increasing the number of iterations. Fourth, as shown in Figure 3.3, the IFTA is only applicable for generating the phase-only hologram of a planar image that is parallel to the hologram plane. The subsequent parts of this section will demonstrate a number of methods for improving the performance of IFTA.

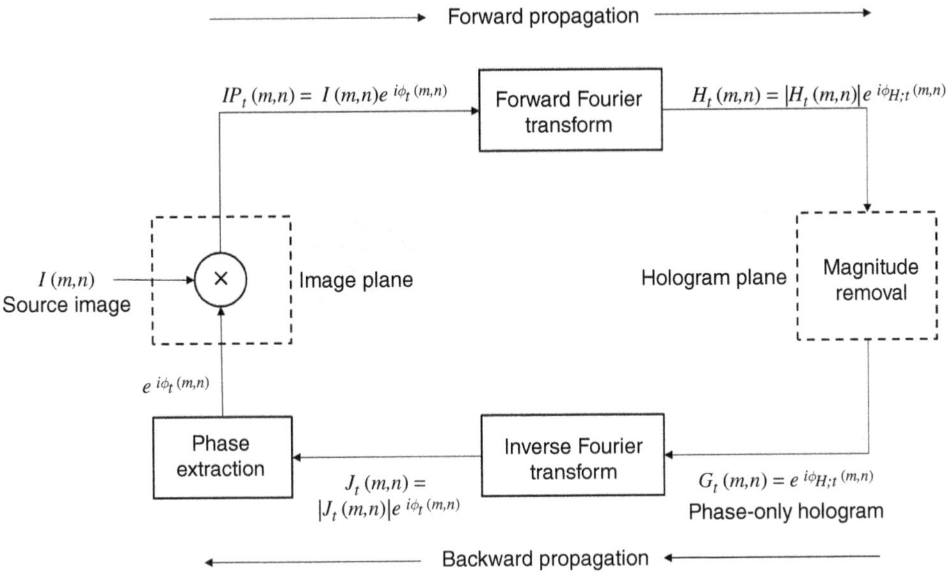

Figure 3.3 The iterative Fourier transform algorithm (IFTA) for generating a phase-only Fourier hologram.

Simulation of the IFTA for generating the Fresnel hologram of a single-depth image is illustrated with the MATLAB code shown in Section 3.4.1. The object space adopted in the simulation is the single-depth image "Apple," shown in Figure 3.4(a), which is located at an axial distance of 0.08 m from the hologram. The image is converted into a phase-only Fresnel hologram with the IFTA method. The reconstructed images on the focused plane of the phase-only holograms that are generated with one, three, and five iterations of the IFTA are shown in Figure 3.4(b–d). In the reconstructed images, only the region that is occupied by the image of the apple is shown. The reconstructed images are similar to the source image, and contaminated with a certain amount of random noise. With more iterations, the noise level is lowered and the reconstructed images become clearer.

3.2.2 Enhanced IFTA: Mixed-Region Amplitude Freedom Method

In Section 3.2.1, the reconstructed image of a phase-only hologram generated with the IFTA suffers noise contamination and slight distortion. The noise signal is caused by the removal of the magnitude component in the hologram, and although part of it can be absorbed into the phase term $e^{i\phi_T(m,n)}$ through the IFTA, the remaining ones will be distributed rather uniformly as random noise in the reconstructed image. In other words, every part of the reconstructed image will be added with a certain amount of random noise. After a certain number of iterations, there will not be further improvement of the noisy appearance of the reconstructed image. To overcome this problem, an enhanced IFTA method known as "mixed-region amplitude freedom" (MRAF) was proposed in [10]. The method was intended for designing

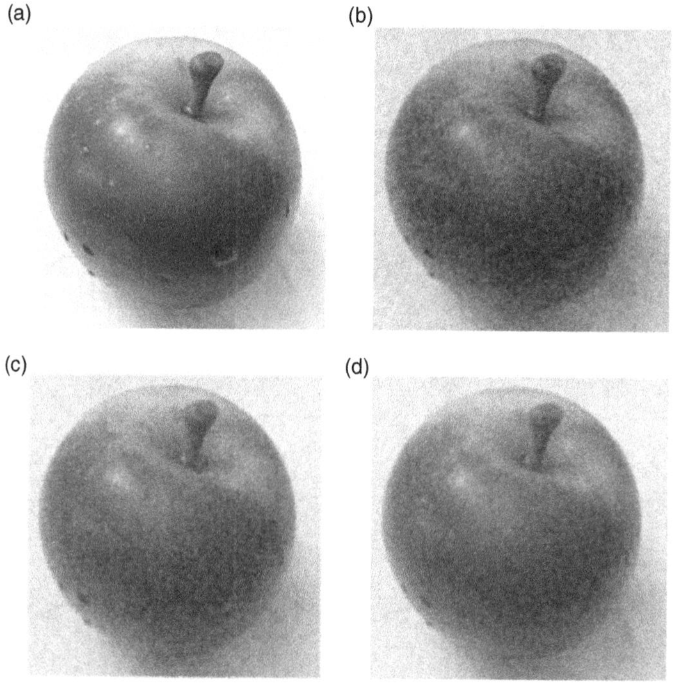

Figure 3.4 (a) Source image "Apple." (b–d) Reconstructed image from a phase-only Fresnel hologram that is generated with one, three, and five round(s) of IFTA, respectively.

holographic atom traps, but it can also be applied in generating phase-only Fourier or Fresnel holograms. In essence, the source image of the object is housed inside a new image of a larger size, so that the majority of noise resulting from the IFTA can be dispatched to the region that is not occupied by the source image. Details of the MRAF method are shown in Figure 3.5.

The source image $I(m,n)$ is first moved to the center of a larger canvas, resulting in a bigger image $I_E(m,n)$. The part that is occupied by $I(m,n)$ (i.e., the dotted region) and the remaining areas on $I_E(m,n)$ are labeled as the "signal region" S and the "noise region," respectively. Initially, at time $t=0$, the pixels of the phase-only hologram $G_{E;t=0}(m,n)$ are assigned random values within the range $[-\pi,\pi]$.

Next, the phase-only hologram is inverse transformed to the reconstructed image $J_{E;t}(m,n) = |J_{E;t}(m,n)|e^{i\phi_t(m,n)}$. On the image plane, the image $IP_{E;t}(m,n)$ is obtained by applying an amplitude constraint on $J_{E;t}(m,n)$:

$$IP_{E;t}(m,n) = \begin{cases} I(m,n)e^{i\phi_t(m,n)} & if\,(m,n)\in\ S \\ J_t(m,n)e^{i\phi_t(m,n)} & \text{otherwise} \end{cases}. \tag{3.3}$$

Equation (3.3) shows that the amplitude constraint enforces the signal region of the reconstructed image to duplicate the source image, whereas in the area covered by the noise region the reconstructed image is not altered. Grossly speaking, the noise region is not subject to amplitude constraint and can be altered freely in the iterative process.

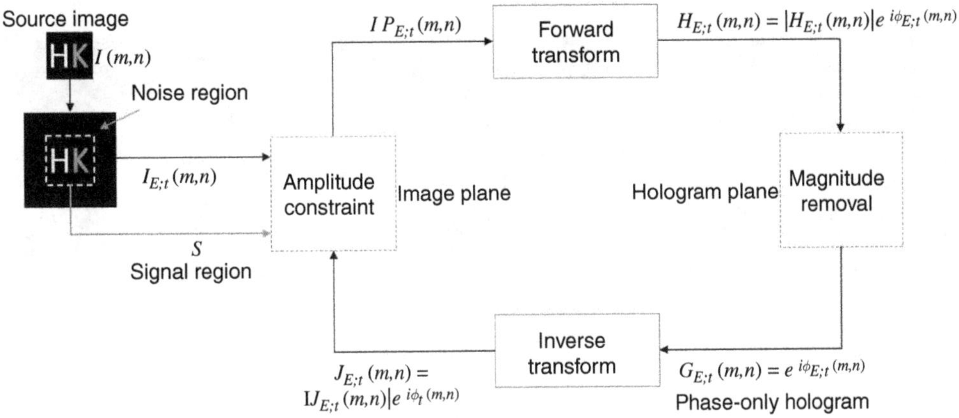

Figure 3.5 The "mixed-region amplitude freedom" (MRAF) for generating a phase-only Fresnel hologram, with the incorporation of the signal and the noise regions.

As the noise region is separated from the signal region containing the source image, it can be used to accommodate part of the distortions that are caused by removing the magnitude component of the hologram.

The amplitude constraint image is then forward transform to its hologram, and the phase component is retained as the phase-only hologram $G_{E;t}(m,n)$, completing the current round of iteration. The above process is repeated until certain criteria, such as reaching the maximum number of iterations, have been met. In each round of iteration, the image within the signal region S is preserved, while the signal in the noise region will be altered, progressively shifting the noise signal from the signal region to the noise region.

Simulation of the MRAF with the incorporation of the signal and noise regions is conducted with the program given in Section 3.4.2. The reconstructed images of the phase-only holograms that are generated with one, three, and five iterations of the MRAF method are shown in Figure 3.6(a–c). The reconstructed image is composed of the image of the apple in the signal region, together with a noisy background in the noise region. Magnified views of the images in the signal region are shown in Figure 3.6 (e–g), and for comparison the original image "Apple" is shown again in Figure 3.6(d). Compared with the results in Figure 3.4(b–d), it is evident that the noise contamination on the reconstructed images has been reduced.

3.2.3 Noise Reduction with IFTA Multiple Frame Averaging

The quality of a phase-only hologram $H_P(m,n)$ generated with IFTA is dependent on the initial randomization of the phase values of the hologram pixels. Suppose the object space is a single planar image $I(m,n)$ that is parallel to the hologram plane, the reconstructed image $I(m,n)$ of the phase-only hologram can be expressed as

$$J(m,n) = I(m,n) + \aleph, \tag{3.4}$$

Figure 3.6 (a–c) Reconstructed image from a phase-only Fresnel hologram that is generated with one, three, and five rounds of MRAF. (d) Source image "Apple." (e–g) Reconstructed image from a phase-only Fresnel hologram that is generated with one, three, and five round(s) of MRAF.

where $\aleph$ represents the noise contamination of the reconstructed image, which is dependent on the initial condition.

To reduce the noise contamination, IFTA is applied to generate a set of phase-only holograms of the same input image, each initialized with different phase randomization. Each of the holograms is referred to as a sub-hologram. Due to the deviation in initial conditions, the reconstructed images of each sub-hologram will be added with a different noise signal. The expectation of the reconstructed image $J(m,n)$ obtained from averaging the reconstructed images of all the sub-holograms is given by

$$\begin{aligned} E[J(m,n)] &= E[I(m,n)] + E[\aleph] \\ &= I(m,n) + E[\aleph]. \end{aligned} \tag{3.5}$$

If the initial randomization is based on white noise, it is likely that the noise in the result will also be similar to white noise. With a sufficient number of samples, $E[\aleph]$ will be close to a homogeneous background and $E[J(m,n)]$ is practically equivalent to the source image.

The above principle can be applied to reduce the noise signal in an IFTA-generated phase-only hologram, and is outlined as follows. To start with, a set of phase-only sub-holograms representing the same source image is generated with the IFTA based on different initial randomization. Subsequently, the sub-holograms are displayed in rapid succession on an SLM. Due to the persistence of vision of human perception, the sequence of reconstructed images will appear to be merged into a single image. If the number of sub-holograms is sufficiently large, the reconstructed image will be close to the source image.

Simulation of this IFTA multiple frame averaging (IFTA-MFA) noise reduction method is conducted here. Similar to previous simulations, the object space is the single-depth image "Apple" in Figure 3.4(a). In the simulation, two or more phase-only sub-holograms are generated with the IFTA under different initial randomizations of the holograms. The magnitude of the reconstructed images of these sub-holograms on the focused plane are averaged to simulate the persistence-of-vision effect.

The reconstructed images obtained with the averaging of 2–5 frames of phase-only sub-holograms that are generated with one iteration of IFTA are shown in Figures 3.7(a–d). Compared with the results in Figure 3.4, it is evident that the noise contamination of the reconstructed images has been reduced, even though only one iteration has been applied in the generation of each phase-only sub-hologram.

The IFTA-MFA has one major disadvantage. In order to achieve better noise reduction, more sub-hologram frames have to be generated and displayed at high speed on the SLM. As a result, the SLM has to operate at a substantially higher refresh rate. With the assumption that the frame rate of a holographic video clip is at the video rate of 25 hologram frames per second, the video refresh rate for five sub-holograms will be $5 \times 25 = 125$ frames per second.

3.2.4 Generating Phase-Only Hologram of a Multi-Depth Object with IFTA

The IFTA described previously is employed in generating a phase-only hologram of a single-depth image that is parallel to the hologram plane. In practice, object points in a 3-D object space could be residing in different depth planes. A 3-D object space can be represented as a sequence of uniformly spaced layers $\Lambda = [I_k(m,n)]|_{1 \le k \le K}$. Each layer is the intensity of a source image that is parallel to the hologram plane, and contains the pixels that are within the layer. In [11], IFTA was extended to the

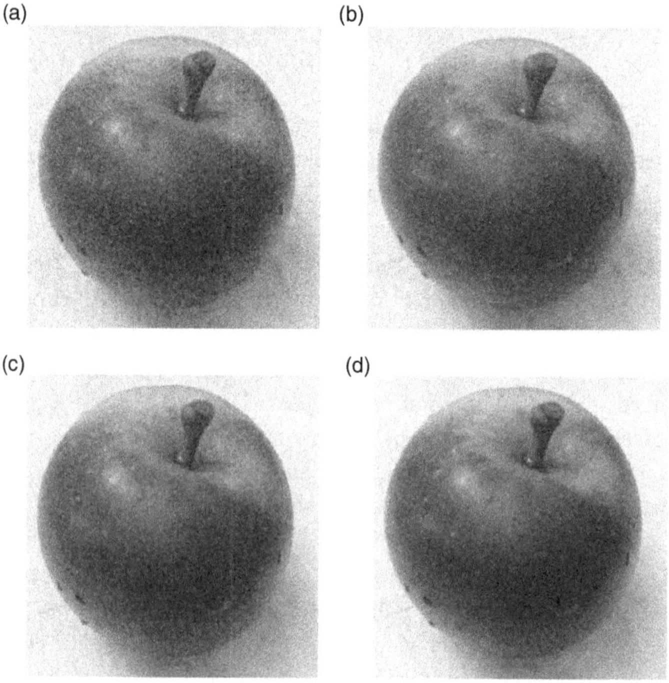

Figure 3.7 Averaging multiple frames of reconstructed images of phase-only Fresnel holograms that are generated with one iteration of IFTA, but under different initial random noise (a)–(d) averaging 2–5 frames, respectively.

generation of an object scene constituted by multiple layers of planar images. The process is depicted in Figure 3.8.

Initially, at time $t = 0$, the phase-only hologram $H_t(m,n)$ is randomly generated, with each hologram pixel assigned a random phase angle in the range $[0, 2\pi)$. The magnitude of all the hologram pixels is set to a constant value of unity, resulting in an initial phase-only hologram. The phase-only hologram is then back propagated to each of the image planes at layer k, where $1 \leq k \leq K$, resulting in a reconstructed image $J_{k;t}(m,n)$:

$$J_{k;t}(m,n) = |J_{k;t}(m,n)| e^{i\phi_{k;t}(m,n)}, \tag{3.6}$$

where $\phi_{k;t}(m,n)$ is the phase component of the reconstructed image $H_t(m,n)$ on the kth focused plane. An amplitude constraint image $IP_{k;t}(m,n)$ is imposed on the image at each reconstruction plane by replacing the magnitude of $J_{k;t}(m,n)$ with the intensity of the corresponding source image $I_k(m,n)$:

$$IP_{k;t}(m,n) = I_k(m,n) e^{i\phi_{k;t}(m,n)}. \tag{3.7}$$

Subsequently, each magnitude constraint image is converted into a sub-hologram with the Fresnel diffraction. The sub-holograms are then summed up into a hologram

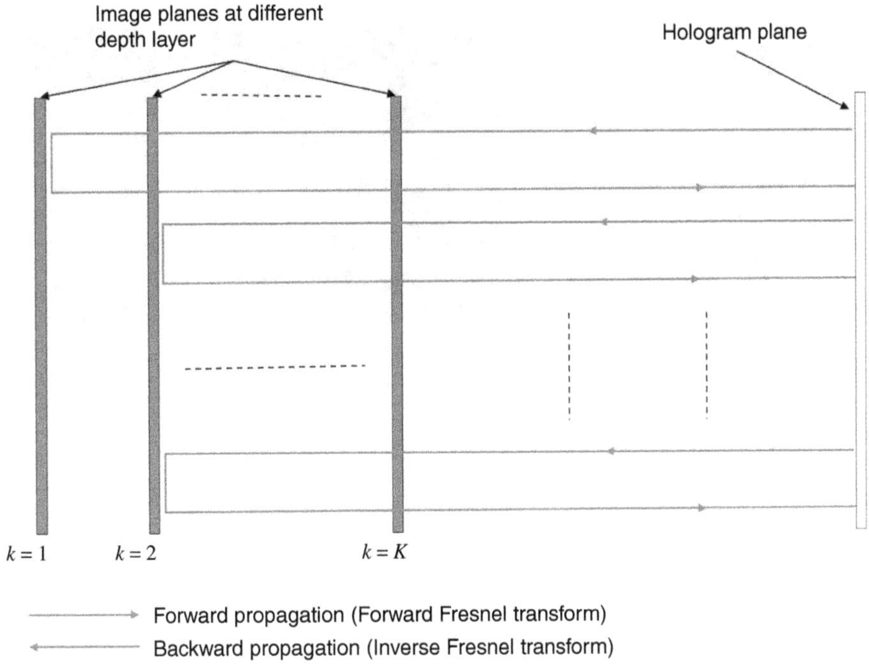

Figure 3.8 The IFTA for generating a phase-only Fresnel hologram of multi-depth image planes.

representing the images of all the depth planes. The iteration index t is advanced by 1, and the phase component of the hologram is extracted to be the phase-only hologram for the current round of IFTA. The process is repeated until the termination criteria are met.

Simulation of the multiple-plane IFTA is conducted. The object space is the double-depth image "HK1" shown in Figure 1.12, comprising a left and a right side, located at 0.07 m and 0.08 m from the hologram, respectively. After conducting five iterations, the phase-only hologram for the double-depth image is generated as shown in Figure 3.9(a). In the figure the pixel values of the phase-only hologram, which has a range of $[0, 2\pi)$, are normalized to $[0, 1)$, respectively. The reconstructed images of the phase-only hologram at the pair of focused planes (0.07 m and 0.08 m) are shown in Figures 3.9(b) and 3.9(c). In each case, the character that resides on the focused plane is reconstructed sharply, with its counterpart appearing as a defocused image.

3.3 Non-iterative Method for Generating Phase-Only Hologram

Although the IFTA is effective in generating a phase-only hologram of a 3-D object image, it has three major disadvantages. First, as the IFTA involves multiple rounds of forward and backward propagations, the computation is rather lengthy, especially if the object space is composed of a large number of depth layers. Second, it is uncertain how many iterations will be required to generate a phase-only hologram that can lead to a faithful reconstruction of the source image. In general, the larger the number of

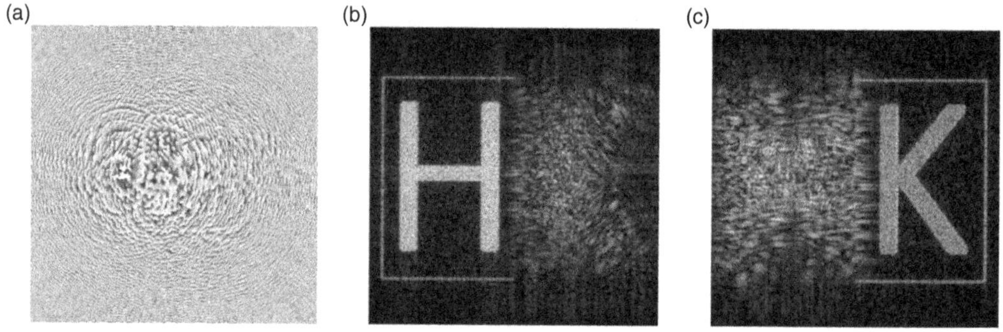

Figure 3.9 (a) Phase-only hologram of the double-depth image in Figure 1.12, generated with five iterations of the multiple-depth IFTA method, (b, c) Reconstructed images from the phase-only Fresnel hologram on the focused plane at 0.07 m and 0.08 m, respectively.

iterations, the better will be the result, but the exact number of iterations to reach a target performance is often dependent on the size and the content of the source image. Third, the reconstructed image of the phase-only hologram generated by the IFTA is contaminated with noise. Although the noise level can be reduced with the incorporation of the signal and noise regions, the noise will be distributed to other parts of the image plane, and some kind of mask is required to block them from sight. In addition, the size of the image has to be enlarged to accommodate the noise window. This will further increase the computation loading as more arithmetic operations are required in the Fresnel transform of a larger image.

The following subsections present a couple of methods that are employed to generate phase-only holograms in a non-iterative manner.

3.3.1 Random Noise Addition

A hologram can be completely represented by its phase component if the magnitude is constant throughout the hologram. In general, the magnitude distribution of hologram pixels is inhomogeneous as it is dependent on the object image. Heavy distortion will be imposed if the magnitude component, which contains essential information on the object image, is removed. One viable solution to overcome this problem is to place a diffuser between the object image and the hologram, so that a light beam of every object point passing through it will be scattered evenly to the entire hologram plane. It can be envisaged that the magnitude distribution of the hologram will become more or less uniform, as the energy of the optical wave at different pixel locations is contributed by the same set of object points. This is, of course, an ideal scenario, but there are bound to be some variations in the scattering of the light wave toward different directions for each object point. Nevertheless, the magnitude distribution will be flattened to an extent that is almost homogeneous.

The insertion of a diffuser can be simulated by adding random phase noise (which is sometimes referred to as the random phase mask) to the intensity image of an object scene.

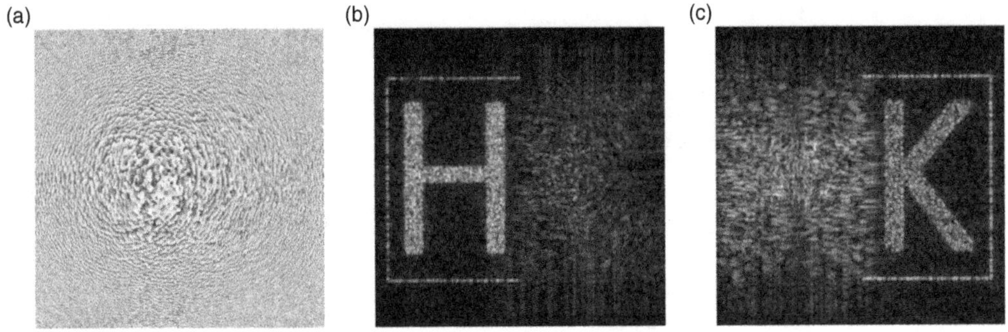

Figure 3.10 (a) Phase-only hologram of a double-depth image, generated with the noise-addition method. (b, c) Reconstructed image of the phase-only Fresnel hologram on the focused plane at 0.07 m and 0.08 m, respectively.

Suppose $I(m,n)$ and $\aleph(m,n)$ denote the intensity image of an object scene and the random phase mask, respectively; the noise-added image is given by

$$I_N(m,n) = I(m,n) \times \aleph(m,n) = I(m,n) \times exp[i\varphi(m,n)], \tag{3.8}$$

where $\varphi(m,n)$ is a 2-D array of random values in the range $[0, 2\pi)$. Subsequently, the noise-added image $I_N(m,n)$ is converted into a hologram with the Fresnel transform.

Simulation of the noise-addition method in generating a phase-only hologram is given in Section 3.4.3. The object space is the double-depth image "HK1" that is shown in Figure 1.12. The noise-addition method is applied to generate the phase-only hologram, as shown in Figure 3.10(a). The reconstructed images of the phase-only hologram on the two depth planes are shown in Figures 3.10(b) and 3.10(c). The images of both characters can be reconstructed from the hologram. However, the reconstructed images are quite heavily contaminated with noise, and some blurriness at the edges is exhibited. In the remaining part of this subsection, two approaches for improving the visual quality of the reconstructed images are introduced.

3.3.2 Edge-Enhanced Noise-Addition Method

Apparently the heavy noise contamination in the reconstructed image shown in Figures 3.10(b) and 3.10 (c) is caused by the addition of a random phase mask to the source image. Prominent damage is also noted around the lines and edges, which are broken in many places. In view of this, a method has been proposed in [12] for improving the visual quality of the reconstructed image. The rationale behind the method is that, as the lines and edge patterns are abrupt changes in the image, they will lead to strong scattering of the optical wave. As such, it is unnecessary to further impose a noise signal on them to produce the scattering effect. The edge-enhanced noise-addition method can be divided into three stages, as shown in Figure 3.11. For simplicity of explanation, we assume that the object space only comprises a single-depth plane with an intensity image $I(m,n)$, but this can be easily extended to multiple-depth planes.

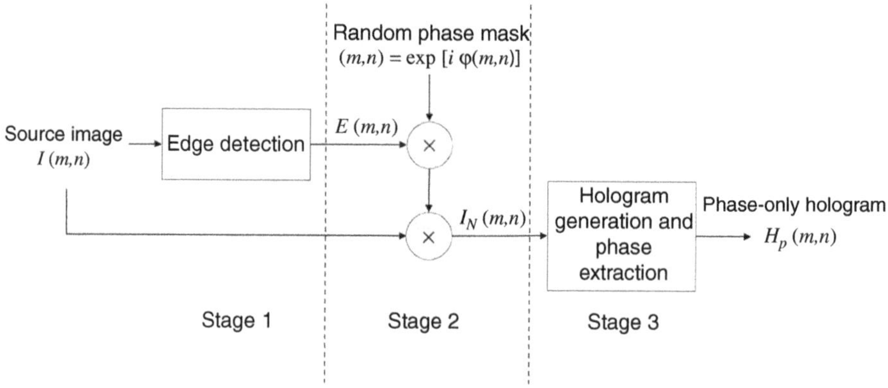

Figure 3.11 Edge-preserved noise-addition method for generating a phase-only hologram.

In the first stage, the edge regions in the intensity image are extracted. Morphological dilation is applied to expand the coverage of the edge regions in the edge image, to ensure that all the lines and edges are included. The result is an edge image, with pixels assigned values of 0 for edge points and 1 for non-edge points. Next, in the second stage, pixels in $I(m,n)$ that are not classified as edge points in the edge image are added with random phase noise as

$$I_N(m,n) = I(m,n) \times \aleph(m,n) \times E(m,n). \tag{3.9}$$

For non-edge points $E(m,n)$ is assigned a value of unity; otherwise it is set to the conjugate of $\aleph(m,n)$. Applying Eq. (3.9), only non-edge pixels of the source image are added with random noise. In the third stage, the noise-added image is converted into a hologram, and the phase component is retained as a phase-only hologram.

A simulation is conducted on the edge-preserved noise-addition method based on the MATLAB code in Section 3.4.3. The object space is the double-depth image in Figure 1.12. Edge detection is first performed on the intensity of the source image, and random phase noise is only added to the non-edge regions. After executing the simulation program, the edge pattern of the source image is shown in Figure 3.12(a), with black and white representing the edge and the non-edge pixels, respectively. The phase-only hologram of the double-depth image "HK1" is generated and shown in Figure 3.12(b), and the reconstructed images of the phase-only hologram on the two focused planes are shown in Figures 3.12(c) and 3.12(d). It can be seen that the images of both characters H and K are well preserved on its focused plane. In addition, the lines and edges in these two figures are much sharper and clearer compared with the ones in Figures 3.10(b) and 3.10(c).

3.3.3 One-Step-Phase-Retrieval

One-step-phase-retrieval (OSPR) [13,14] was developed by Buckley for fast generation of phase-only holograms in a non-iterative manner. The method is based on the principle of noise addition, and is depicted in Figure 3.13. First, a set of noise-added

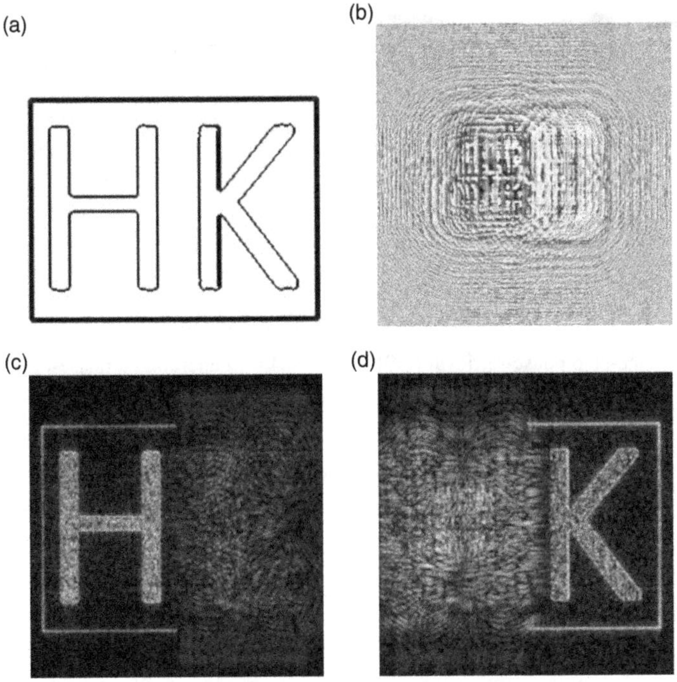

Figure 3.12 (a) Edge pattern of the intensity of the double-depth source image. (b) Phase-only hologram of a double-depth image, generated with the noise-addition method. (c, d) Reconstructed image of the phase-only Fresnel hologram on focused planes at 0.07 m and 0.08 m, respectively.

images is generated by applying Eq. (3.8) to the intensity image $I(m,n)$ with K different random phase masks. As mentioned previously, the random phase mask is equivalent to a diffuser for scattering the object waves evenly over the hologram plane. Second, a Fresnel hologram is generated $I_{N;k}(m,n)|_{1\leq k\leq K}$ for each noise-added image, and the magnitude of the pixels is discarded (i.e., assigned a value of unity), and the phase component is retained as a phase-only sub-hologram. The phase-only hologram can be further quantized into a binary representation through thresholding or error diffusion [14]. The reconstructed images of each phase-only hologram will be similar in appearance to the ones in Figures 3.10(b) and 3.10(c), with both characters overlaid with substantial amounts of noise. Finally, the set of phase-only sub-holograms are displayed sequentially at a high frame rate. Due to the persistence of vision of human perception, the sub-holograms – and hence the reconstructed images – will appear to overlap each other. As a result, the noise in the reconstructed images will be averaged out by the eyes and will become less prominent.

Simulation of the OSPR method in generating a phase-only hologram of the double-depth image "HK1" in Figure 1.12. is conducted. Multiple phase-only holograms are generated with the OSPR method with $K=4$ and $K=16$, and the reconstructed images on the pair of depth planes for these two cases are shown in Figures 3.14(a, b)

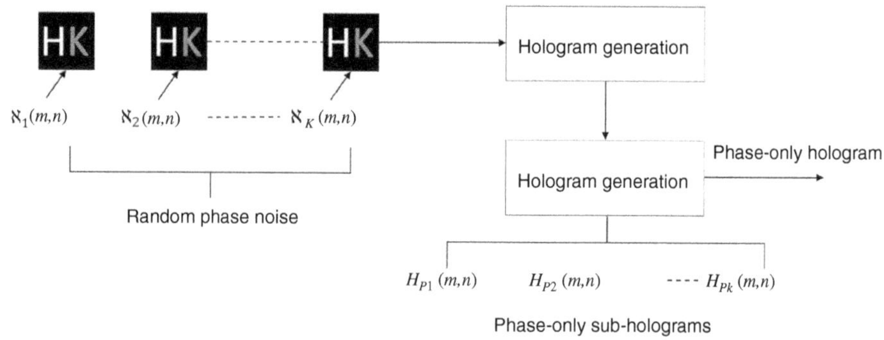

Figure 3.13 One-step phase-retrieval (OSPR) method for generating a phase-only hologram.

and 3.14(c, d), respectively. Compared with the results in Figure 3.10, the noise contamination on the reconstructed images is reduced, and the noise level is lower in the case of $K = 16$.

Although the OSPR method is effective in reducing the noise in the reconstructed image, the computation time is lengthy as multiple phase-only holograms have to be generated. In addition, a fast SLM is required to display the multiple phase-only holograms at a high frame rate so that their reconstructed images can be merged through the persistence of vision of human perception.

3.3.4 Patterned Phase-Only Hologram

From the methods described in Sections 3.3.1–3.3.3, it can be seen that the noise-addition method is a very fast way of generating a phase-only hologram, as it does not impose extra computation loading on the CGH process other than adding random phase noise to the intensity image of the object scene. However, the reconstructed images of a phase-only hologram generated with the noise-addition method are heavily contaminated with noise. In OSPR, the noise level is lowered by averaging multiple sub-holograms through persistence of vision. However, the computation time is substantially lengthened.

A method to overcome the abovementioned problems, hereafter referred to as the pattern phase method, was proposed in [15]. Similar to the noise-addition method, a diffuser in the form of a phase mask is added to scatter the light waves of each object point, so that the magnitude of the wavefront will be more or less uniform on the hologram plane. Instead of a totally random phase distribution, repetitive square blocks of phase patterns are employed to form the phase mask. The size of a square block is $\tau \times \tau$, and the phase of each pixel within it, denoted by $\theta_{pp}(m,n)|_{0\leq\ m;n<\tau}$, is a random value in the range $[0, 2\pi)$. After imposing the phase mask, the intensity image becomes

$$I_{pp}(m,n) = I(m,n) \times \theta_{pp}\Big(\mathrm{mod}(m,\tau), \mathrm{mod}(\tau,n)\Big), \tag{3.10}$$

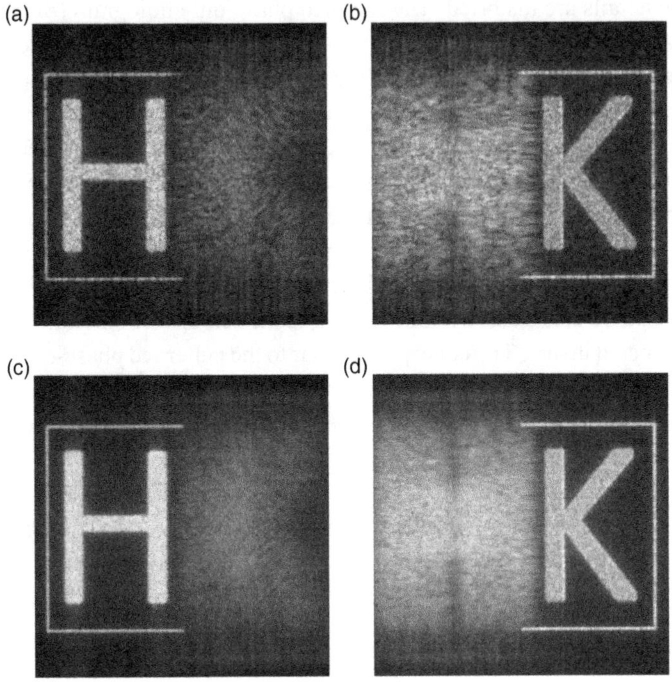

Figure 3.14 (a, b) Reconstructed image from a phase-only Fresnel hologram that is generated with the OSPR method, based on the averaging of four holograms, on the focused planes at 0.07 m and 0.08 m. (c, d) Reconstructed image from a phase-only Fresnel hologram that is generated with the OSPR method, based on the averaging of 16 holograms, on the focused planes at 0.07 m and 0.08 m.

where $a, b \in \mathbb{Z}$, and $\mathrm{mod}(a, b)$ is the modulus operator that returns the remainder of a/b. Subsequently, a Fresnel hologram is generated and the phase component is retained as a patterned phase-only hologram.

The MATLAB code for generating the patterned phase-only hologram for the double-depth image "HK1" in Figure 1.12 is listed in Section 3.4.4. Simulations are conducted with block size $\tau = 4$ and $\tau = 8$. The patterned phase-only holograms for these two cases are shown in Figures 3.15(a) and 3.15(b). The reconstructed images on the pair of focused planes are shown in Figure 3.15(c, d) for $\tau = 4$ and Figure 3.15(e, f) for $\tau = 8$. From the reconstructed images, it can be seen that the smooth-shaded regions are preserved favorably, but masked with a coarse pattern that is repeated throughout the reconstructed image, resulting in a pixelated appearance. It is noted that with a smaller block size, the pixelated effect is reduced and the edges are sharper at the expense of lowering the brightness and uniformity in the shaded areas.

3.3.5 Sampled Phase-Only Hologram

Although the patterned phase-only hologram is effective in preserving the reconstructed image with favorable quality, a pixelated appearance is imposed and some of the fine details are distorted. The sampled phase-only hologram (SPOH) [16] is an alternative method for generating a phase-only hologram that retains more fine details of the source image. In this approach, the intensity distribution of the source image is downsampled with a uniformly grid-cross lattice before it is converted into a hologram. The downsampling process is equivalent to overlaying a dense diffraction grating onto the object scene, creating a regular pattern of closely spaced discontinuities on the surface of the object. All of these discontinuities generate strong diffracted waves in all directions, so that their accumulated intensity distribution will be more or less homogeneous on the hologram plane. As such, the magnitude component of the hologram can be discarded without causing too much distortion on the hologram and its reconstructed image. The concept is similar to the patterned phase-only hologram or the noise-addition method, but in an SPOH there is no external added noise to contaminate the source image. Details of the generation of the SPOH are shown in Figure 3.16(a), and outlined as follows.

The intensity distribution $I(m,n)$ of a 3-D object surface is downsampled with a grid-cross lattice, resulting in a downsampled image $I_D(m,n)$. The grid-cross lattice comprises of uniformly spaced horizontal, vertical, and diagonal lines, represented by a binary image $S(m,n)$ with pixel values of 1 and 0 corresponding to sampled and

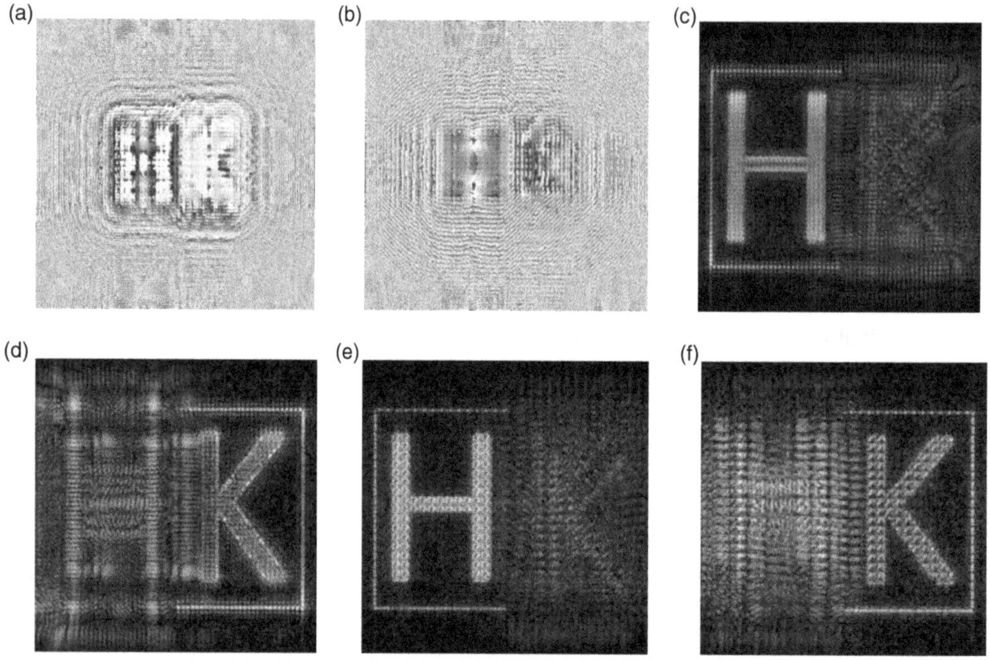

Figure 3.15 (a) Patterned phase holograms of the image "HK1" with block size $\tau = 4$. (b) Patterned phase holograms of the image "HK1" with block size $\tau = 8$. (c, d) Reconstructed image from the patterned phase-only hologram in Figure 3.15(a) on the focused planes at 0.07 m and 0.08 m. (e, f) Reconstructed image from the patterned phase-only hologram in Figure 3.15(b) on the focused planes at 0.07 m and 0.08 m.

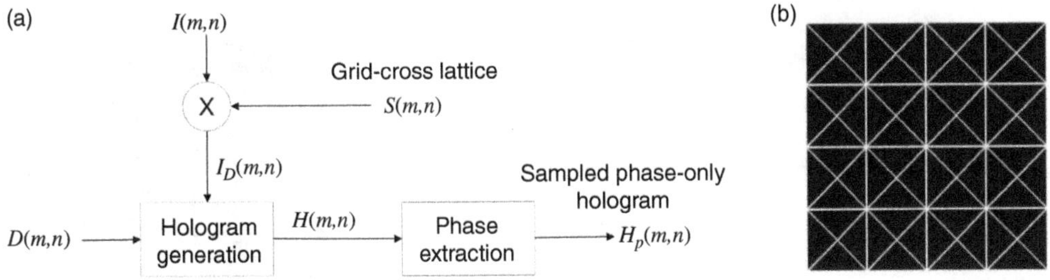

Figure 3.16 (a) The process of generating an SPOH. (b) A small section of the grid-cross downsampling lattice.

non-sampled points, respectively. A small section of the grid-cross downsampling lattice is shown in Figure 3.16(b).

After downsampling with the grid-cross lattice, only pixels along the horizontal, vertical, and the up-/down-diagonal lines will be retained, while the rest of them will be removed. Mathematically, $S(m,n)$ can be expressed as the union of four downsampling sub-lattices, each corresponding to one set of the downsampling lines:

$$S(m,n) = \bigcup_{k=1}^{4} S_k(m,n), \tag{3.11}$$

where $S_1(m,n)$, $S_2(m,n)$, $S_3(m,n)$, and $S_4(m,n)$ are the sets of horizontal, vertical, up-diagonal, and down-diagonal sampling lines. Let τ denote the downsampling factor and $\%$ denote the modulus operator:

$$S_1(m,n) = \begin{cases} 1 & \textit{if } m\%\tau = 0 \\ 0 & \text{otherwise} \end{cases}, \tag{3.12a}$$

$$S_2(m,n) = \begin{cases} 1 & \textit{if } n\%\tau = 0 \\ 0 & \text{otherwise} \end{cases}, \tag{3.12b}$$

$$S_3(m,n) = \begin{cases} 1 & \textit{if } (m\%\tau) = (n\%\tau) \\ 0 & \text{otherwise} \end{cases}, \tag{3.12c}$$

$$S_4(m,n) = \begin{cases} 1 & \textit{if } (m\%\tau) = \tau - (n\%\tau) \\ 0 & \text{otherwise} \end{cases}. \tag{3.12d}$$

The downsampled image is obtained from the product of the intensity distribution of the 3-D object surface and the downsampling lattice:

$$I_d(m,n) = I(m,n) \times S(m,n). \tag{3.13}$$

Next, the downsampled image is converted into a complex-valued hologram $H(m,n)$, and the phase component is retained as the phase-only hologram $H_P(m,n)$.

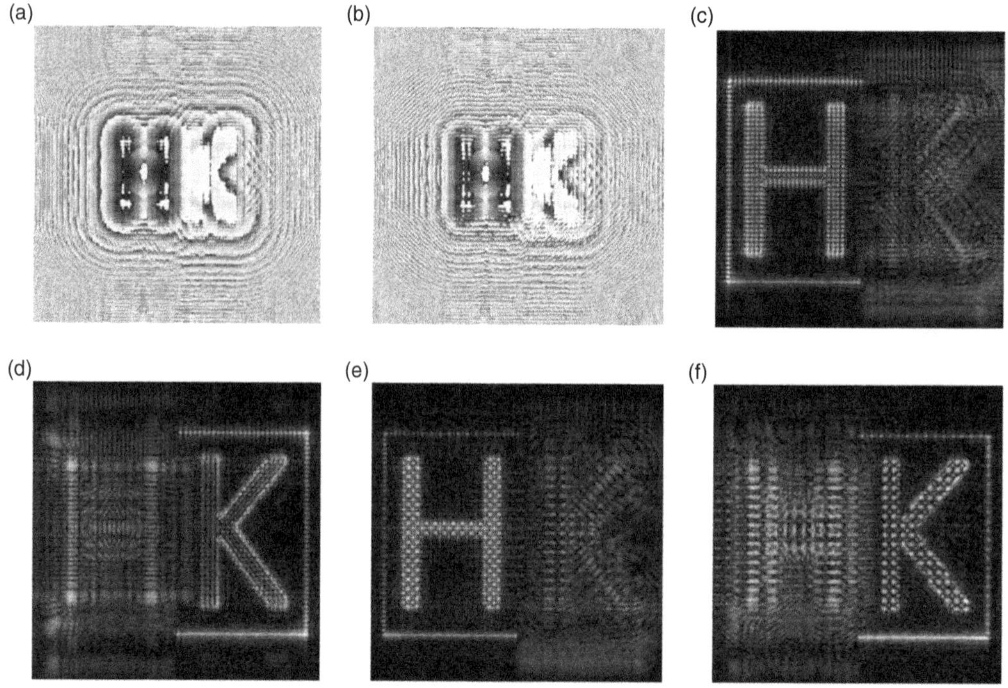

Figure 3.17 (a) Sampled phase holograms of the image "HK1" with block size $\tau = 4$. (b) Sampled phase holograms of the image "HK1" with block size $\tau = 8$. (c, d) Reconstructed image from the SPOH in Figure 3.17(a) on the focused planes at 0.07 m and 0.08 m. (e, f) Reconstructed image from the SPOH in Figure 3.17(b) on the focused planes at 0.07 m and 0.08 m.

Modifying the MATLAB codes in Section 3.4.4, a simulation is conducted for generating an SPOH for the double-depth image "HK1" in Figure 1.12. The SPOHs and the reconstructed images on the pair of focused planes for sampling factors $\tau = 4$ and $\tau = 8$ are shown in Figures 3.17(a, b), Figure 3.17(c, d), and Figure 3.17(e, f). In both sets of examples, with a smaller sampling factor τ, the brightness and uniformity of the shaded areas are lowered, but the lines and edges are preserved more favorably. The other way round is noted for larger values of τ.

3.3.6 Edge-Enhanced Sampled Phase-Only Hologram

In the Section 3.35, the sampling factor must be sufficiently large in order to preserve the shaded regions in the reconstructed image. However, a larger sampling factor also leads to more severe fragmentation of the lines and edges. In this section, a variation of the SPOH generation method for overcoming this dilemma is described. The method, which is referred to as edge-enhanced sampled phase-only hologram (EESPOH) [17], is outlined here.

Similar to the generation of an SPOH, the intensity component of the source image, denoted by $I(m, n)$, is downsampled with a lattice $S_A(m, n)$ as

$$I_d(m,n) = I(m,n) \times S_A(m,n). \tag{3.14}$$

The downsampling lattice is a union of two sub-lattices,

$$S_A(m,n) = S(m,n) \bigcup S_E(m,n). \tag{3.15}$$

The first downsampling sub-lattice $S(m,n)$ is the grid-cross pattern represented by Eqs. (3.11) and (3.12). The second sub-lattice $S_E(m,n)$ corresponds to the lines and edges of the source image, and is obtained with edge detection. First, a Laplacian kernel $L(m,n)$ is applied to convolve with $I(m,n)$, resulting in an interim image given by

$$I_H(m,n) = I(m,n)^* L_P(m,n), \tag{3.16}$$

where $L_P(m,n)|_{-1\le m,n\le 1} = \begin{cases} 1 & m=n=0. \\ \dfrac{-1}{8} & \text{otherwise} \end{cases}$

The Laplacian kernel is a high-pass filter operator that attenuates the low-frequency component of the original image. Next, the interim image is converted into the second downsampling sub-lattice with thresholding:

$$S_E(m,n) = \begin{cases} 1 & I_H(m,n) \ge T \\ 0 & \text{otherwise} \end{cases}. \tag{3.17}$$

Applying Eq. (3.15), the downsampled image $I_d(m,n)$ is extracted from the lines and edges of $I(m,n)$, as well as the pixels that are sampled by the uniform grid-cross pattern. Subsequently, the downsampled image is converted into a digital Fresnel hologram, and the phase component is retained as a phase-only hologram. In the reconstructed image, the shaded areas are preserved favorably due to the downsampling of the intensity image. At the same time, the lines and edges are degraded to a much lesser extent than in the SPOH method, as they are not fragmented with downsampling in the first place.

3.3.7 Complementary Sampled Phase-Only Hologram

One of the major disadvantages of the SPOH is that the source image has to be downsampled with a grid-cross lattice prior to the generation of the hologram. As a result, both the source image and the reconstructed image of an SPOH are sparse, leading to fragmented lines and edges, and empty voids in the shaded areas. Although the EESPOH method can improve the quality of the lines and edges, the sparsity of the shaded areas is not rectified. The empty voids in the source image can be reduced with a smaller downsampling interval. However, as shown in Section 3.3.5, the downsampling interval must be larger than a certain threshold, below which the shaded regions in the reconstructed image will be heavily distorted.

A viable solution to overcome the sparsity problem in an SPOH is a method known as the complementary sampled phase-only hologram (CSPOH), as proposed in [18]. In this method, a pair of SPOHs are generated from the same source image, but based on two different downsampling lattices, L_1 and L_2. The first lattice L_1 is a grid-cross sampling lines, similar to the one in the parent SPOH method. The second lattice L_2 is a lattice composed of sampling points that are complementary to the first one. Each of the SPOHs is taken to reconstruct a sampled version of the source image. As the pair of SPOHs are swapping at a high frame rate, to an observer the missing pixels in one reconstructed image will be filled by those in the other one due to persistence of vision of human eyes.

3.3.8 Binary Phase-Only Hologram

In some applications, it is preferable to generate a binary phase-only hologram (BPOH) instead of a continuous tone phase-only hologram. A binary hologram has smaller data size and is easier to produce with a hologram printer. A BPOH can be obtained by quantizing a continuous phase-only hologram into a bi-level representation. For example, assigning 0 and π radians to the phase angle within the range $[0, \pi)$, and $[\pi, 2\pi)$, respectively. However, this will lead to heavy distortion on the hologram and its reconstructed image.

Direct binary search (DBS) is an iterative method for generating a BPOH [19] through an iterative optimization process. Its principles of operation can be applied to derive binary Fourier, amplitude-only, or phase-only holograms. The original complex-valued hologram is denoted by $H(m,n)$, representing a source image $I(m,n)$. The goal is to generate a BPOH so that its reconstructed image is similar to a reference image, which is either the source image or the reconstructed image of $H(m,n)$ if $I(m,n)$ is not available. For simplicity of explanation, it is assumed that the source image is known. The binary hologram at the kth iteration is given by $B_k(m,n)$. Initially at $k=0$, pixels in the binary hologram are assigned random values of either 1 or 0, and the reconstructed image on the focused plane is computed as

$$I_{B;k}(m,n) = B_k(m,n) * h(m,n;z_0), \tag{3.18}$$

where $h(m,n;z_0)$ is the free space impulse response. In the execution of the DBS, pixels in the binary hologram are scanned lexicographically along a row-by-row, and left-to-right direction. When each pixel is visited, its value will be inverted and the index of the iteration k is increased by 1. The reconstructed image of the updated binary hologram on the focused plane is computed with Eq. (3.14).

A pair of error quantities corresponding to the mean square errors between the source image $I(m,n)$ and the reconstructed images at the current iteration $I_{B;k}(m,n)$ and the previous iteration $I_{B;k-1}(m,n)$ are determined as

$$E_{k-1} = MSE[I(m,n), I_{B;k-1}(m,n)], \tag{3.19}$$

and

$$E_k = MSE[I(m,n), I_{B;k}(m,n)], \tag{3.20}$$

where $MSE[A, B]$ denotes the mean-square-error of A and B. If $E_k < E_{k-1}$, the updated binary hologram is retained; otherwise it will be reverted to the form in the previous iteration – that is, $B_k(m, n) = B_{k-1}(m, n)$. The process is repeated after all the binary pixels have been evaluated. The DBS is in fact a systematic trial-and-error optimization process, whereby the decision on the binary value of each hologram pixel is based on one that results in a smaller amount of error.

There are two major disadvantages of the DBS algorithm. First, in the evaluation of the binary value of each pixel, Eq. (3.14) has to be applied to obtain the reconstructed image. As every pixel in the binary hologram has to be evaluated, the computation time is extremely long. Second, the significance of each hologram pixel is dependent on the order of its visit. Hologram pixels that are visited at the earlier stage of the search are reflecting more of their own contributions (as most of the pixels in the hologram are not tested yet). At the later stage of the process, determination of the binary state of each hologram pixel will be highly dependent on the pixels that have been evaluated. While the second problem can be alleviated by conducted multiple rounds of DBS, the computation time will be prolonged, hence escalating the first problem.

A more effective method for overcoming the abovementioned problems were proposed in [20]. As pointed out in that paper, the quality of the reconstructed image of the binary hologram can be improved simply by visiting the hologram pixels in a random instead of lexicographic manner. Furthermore, a block partitioned strategy had been adopted to reduce the computation time. Instead of optimizing the binary hologram as a whole, the original complex-valued hologram is partitioned into non-overlapping blocks. Initially, each block is assigned random binary pixels. A set of sub-holograms is formed, each composed of one of the partitioned blocks, with the rest of the area identical to the source hologram. Next, DBS is applied to optimize the partitioned block in each sub-hologram. In this approach, optimizing the partitioned blocks can be conducted concurrently, as the DBS process conducted in each sub-hologram is independent of the others. This permits the use of parallel computing devices (such as graphic processing units) to speed up the computation.

3.4 MATLAB Simulation

This section includes the MATLAB source codes that are employed to conduct the simulations of the methods reported in this chapter. A summary of the simulation programs (each identified by the name of its main function call), and their correspondence to sections in the chapter, are listed in Table 3.1.

Two different images are used in the simulations. The first one is the double-depth image “HK1” shown in Figure 1.12. The image is evenly divided into a left half-plane and a right half-plane containing the characters H and K, respectively. The left half-plane is located at a distance of 0.07 m from the hologram plane, while the right half-plane is placed farther away at 0.08 m. The second image is “Apple,” which is located at an axial distance of 0.08 m from the hologram.

Table 3.1 Summary of simulation programs in this chapter

Function name	Description of simulation	Associated sections
IFTA	Computer-generated hologram of a single-depth image with the IFTA	3.2.1
MRAF	Computer-generated hologram of a single-depth image with the MRAF and the incorporation of the signal and the noise windows	3.2.2
NA	Computer-generated hologram of a double-depth image with the noise-addition method	3.3.1
PPOH	Computer-generated hologram of a double-depth image with the patterned phase-only hologram method	3.3.4

The image "Apple" is adopted in the simulation of IFTA and MRAF. The double-depth image "HK1" is used in the simulation of the noise-addition and the PPOH methods.

3.4.1 Simulation of Generating a Phase-Only Hologram with IFTA

In this simulation, with the source code shown in Table 3.2, a phase-only hologram of a source image is generated with the IFTA method. To begin with, the main function "IFTA," is called to load the source image file. The user is prompted to enter the number of iterations to be conducted with the IFTA. Next, the image and the number of iterations specified by the user are passed to the function "IFTA_iter." The size of the hologram is set to be four times that of the image. Upon completion of the IFTA, a phase-only hologram of the source image is generated, and the reconstructed image at the focused plane will be computed from the phase-only hologram with the function "Holo_recon." Subsequently, the reconstructed image is displayed and saved.

3.4.2 Simulation of Generating a Phase-Only Hologram with the MRAF

The source code for the simulation is shown in Table 3.3. In this simulation, a phase-only hologram of the single-depth image "Apple" is generated with the MRAF method. The image is located at an axial distance of 0.08 m from the hologram. To begin with, the main function "MRAF" is called to load the source image file. Next, the function "MRAF_iter" is applied to the source image for a number of iterations that is input by the user. The MRAF is similar to IFTA, but the amplitude constraint is only applied to the signal region. Upon completion of the MRAF, a phase-only hologram of the source image is generated, and the reconstructed image at the focused plane will be computed from the phase-only hologram with the function "Holo_recon." Subsequently, the reconstructed image is displayed and saved.

Table 3.2 Simulation on the IFTA method

```
function IFTA()
  close all; clc; warning off
  fname='Apple';
  I=imread(strcat(fname,'.jpg'));
  I=im2double(I);
  psize=6.4e-6;                        %Pixel size of hologram=6.4 µm
  lamda=520e-9;                        %Wavelength=520 nm
  z=0.08;                              %Depth plane of source image at 0.08 m
  iter=input('Enter number of iterations for the IFTA ');
  HP=IFTA_iter(I,z,psize,lamda,iter);
  %Reconstruct image at focused plane from phase-only hologram HP
  RES=mat2gray(abs(Holo_recon(HP,z,psize,lamda)));
  [M N]=size(RES(:,:))                 %Size of hologram
  [Y X]=size(I(:,:))                   %Size of source image I(m,n)
  sy=(M-Y)/2; ey=(M+Y)/2-1;
  sx=(N-X)/2; ex=(N+X)/2-1;
  RES_Corp=RES(sy:ey,sx:ex);           %display image = Two times the size of
                                         %the source image
  imwrite(RES,strcat('RES_IFTA_',num2str(iter),'_UC_',fname,'.jpg'));
  imwrite(RES_Corp,strcat('RES_IFTA_',num2str(iter),'_C_',fname,'.jpg'));
  figure,imshow(RES);                  %Display reconstructed image at z
  title('Reconstructed image of the POH obtained with IFTA');
  figure,imshow(RES_Corp);             %Display cropped reconstructed image
                                         %at z
  title('Reconstructed image, cropped to image window, of the POH obtained with
   IFTA');
  end
function HP=IFTA_iter(I_org,z,psize,lamda,iter)
  I_org=I_org/max(I_org(:));
  [Y X]=size(I_org(:,:));              %Size of source image
  sc=4;                                %Hologram is sc times larger than the
                                         %image
  %Set hologram and new image size to be four times that of original image
  N=sc*Y;
  M=sc*X;
  I=zeros(N,M);
  %Move source image to center of the enlarged image plane
  sy=(N-Y)/2; ey=(N+Y)/2-1;
  sx=(M-X)/2; ex=(M+X)/2-1;
  I(sy:ey,sx:ex)=I_org;
  I0=I;
  H=Holo_gen(I0,z,psize,lamda);        %Generate hologram of source image
  %================================================
  %Conduct multiple (iter) rounds of IFTA
  for k=1:iter
   if k==1                             %Initially, set random values to the
                                         %phase-only hologram
     ang=rand(N,M)*6.2832;
```

Table 3.2 (cont.)

```
    else                                    %Else retain phase values of the
                                              %hologram
      ang=angle(HP);
    end
    HP=cos(ang)+j*sin(ang);
    %Back propagate phase-only hologram to image plane
    J=Holo_recon(HP,z,psize,lamda);
    JA=angle(J);                            %Phase angle of reconstructed image
    IP=I.*(cos(JA)+j*sin(JA));              %Multiply source image with phase angle
                                              %JA
    H=Holo_gen(IP,z,psize,lamda);           %Generate hologram of the image IP
   HP=cos(angle(H))+1i*sin(angle(H));%Extract phase component of hologram
  end
end
function H=Holo_gen(I,z,psize,lamda)
  [N M]=size(I(:,:));
  wn=2*pi/lamda; %Wave number
  %Generate Fresnel zone plate
  z2=z*z;                                   %Square of the depth value z
  [kx ky]=meshgrid(-M/2:M/2-1,-N/2:
  N/2-1);
  kx=kx*psize;                              %Physical horizontal position
  ky=ky*psize;                              %Physical vertical position
  fzp=exp(j*wn*sqrt(kx.^2+ky.^2+z2));
  %Compute hologram of source image
  A=fft2(I);                                %Fourier transform of the hologram
  B=fft2(fzp);                              %Fourier transform of the free space
                                              %impulse response
  H=fftshift(ifft2(A.*B));
end
function IR=Holo_recon(H,z,psize,
  lamda)
  [N M]=size(H(:,:));                       %Size of hologram
  wn=2*pi/lamda;                            %Wave number
  %Generate Fresnel zone plate
  z2=z*z;                                   %Square of the depth value z
  [kx ky]=meshgrid(-M/2:M/2-1,-N/2:
  N/2-1);
  kx=kx*psize;                              %Physical horizontal position
  ky=ky*psize;                              %Physical vertical position
  r=sqrt(kx.^2+ky.^2+z2);
  fzp_c=exp(-j*wn*r)./r;                    %Conjugate of FZP
  A=fft2(H);                                %Fourier transform of the hologram
  B=fft2(fzp_c);                            %Fourier transform of the free space
                                              %impulse response
  IR=fftshift(ifft2(A.*B));                 %Compute reconstructed image
end
```

Table 3.3 Source code for simulating the MRAF method

```
function MRAF()
  close all; clc; warning off
  fname='Apple';
  I=imread(strcat(fname,'.jpg'));
  I=im2double(I);
  psize=6.4e-6;                        %Pixel size of hologram=6.4 µm
  lamda=520e-9;                        %Wavelength=520 nm
  z=0.08;                              %Depth plane of source image at 0.08 m
  iter=input('Enter number of iterations for the MRAF ');
  HP=MRAF_iter(I,z,psize,lamda,iter);
  %Reconstruct image at focused plane from phase-only hologram HP
  RES=mat2gray(abs(Holo_recon(HP,z,psize,lamda)));
  [N M]=size(RES(:,:));                %Size of hologram/reconstructed image
  [Y X]=size(I(:,:));                  %Size of source image I(m,n)
  sy=(N-Y)/2; ey=(N+Y)/2-1;
  sx=(M-X)/2; ex=(M+X)/2-1;
  %Cropped reconstructed image to include the signal region only
  RES_Corp=RES(sy:ey,sx:ex);
  imwrite(RES,strcat('RES_MRAF_UC_',num2str(iter),'_',fname,'.jpg'));
  imwrite(RES_Corp,strcat('RES_MRAF_C_',num2str(iter),'_',fname,'.jpg'));
  figure,imshow(RES);                  %Display reconstructed image at
                                         %focused plane
  title('Reconstructed image of the POH obtained with MRAF');
  figure,imshow(RES_Corp);             %Display reconstructed image at z
  title('Reconstructed image within signal window of the POH obtained with
  MRAF');
end
function HP=MRAF_iter(I_org,z,psize,lamda,iter)
  I_org=I_org/max(I_org(:));
  [Y X]=size(I_org(:,:));              %Size of source image
  sc=4;                                %Hologram is sc times larger than the
                                         %image
  %Set hologram and new image size to be four times that of original image
  N=sc*Y; M=sc*X;
  I=zeros(N,M);
  %Move source image to center of the enlarged image plane
  %The part occupied by the image is the signal region, the rest is the noise
  %region
  sy=(N-Y)/2; ey=(N+Y)/2-1;
  sx=(M-X)/2; ex=(M+X)/2-1;
  I(sy:ey,sx:ex)=I_org(1:Y,1:X);
  IP=zeros(N,M);
  %Conduct multiple rounds of MRAF
  for k=1:iter
   if k==1                             %Initially, set random values to the
                                         %phase-only hologram
    ang=rand(N,M)*2*pi;
```

Table 3.3 (cont.)

```
  else                                           %Else retain phase values of the
                                                   %hologram
   ang=angle(HP);
  end
  HP=cos(ang)+j*sin(ang);
  J=Holo_recon(HP,z,psize,lamda);
  J=J/max(abs(J(:)));
  JA=angle(J);
  IP=J;
  %Only signal region is subject to amplitude constraint
  IP(sy:ey,sx:ex)=I(sy:ey,sx:ex).*(cos(JA(sy:ey,sx:ex))+j*sin(JA(sy:ey,
 sx:ex)));
  H=Holo_gen(IP,z,psize,lamda);                  %Propagate image to hologram
  HP=cos(angle(H))+1i*sin(angle(H));%Preserve phase component only
 end
end
function H=Holo_gen(I,z,psize,lamda)
 [N M]=size(I(:,:));
 wn=2*pi/lamda;                                  %Wave number
 %Generate Fresnel zone plate
 z2=z*z;                                         %Square of the depth value z
 [kx ky]=meshgrid(-M/2:M/2-1,-N/2:
 N/2-1);
 kx=kx*psize;                                    %Physical horizontal position
 ky=ky*psize;                                    %Physical vertical position
 fzp=exp(j*wn*sqrt(kx.^2+ky.^2+z2));
 %Compute hologram of source image
 A=fft2(I);                                      %Fourier transform of the hologram
 B=fft2(fzp);                                    %Fourier transform of the free space
                                                   %impulse response
 H=fftshift(ifft2(A.*B));
end
function IR=Holo_recon(H,z,psize,
  lamda)
 [N M]=size(H(:,:));                             %Size of hologram
 wn=2*pi/lamda;                                  %Wave number
 %Generate Fresnel zone plate
 z2=z*z;                                         %square of the depth value z
 [kx ky]=meshgrid(-M/2:M/2-1,-N/2:
 N/2-1);
 kx=kx*psize;                                    %Physical horizontal position
 ky=ky*psize;                                    %Physical vertical position
 r=sqrt(kx.^2+ky.^2+z2);
 fzp_c=exp(-j*wn*r)./r;                          %Conjugate of FZP
 A=fft2(H);                                      %Fourier transform of the hologram
 B=fft2(fzp_c);                                  %Fourier transform of the free space
                                                   %impulse response
 IR=fftshift(ifft2(A.*B));                       %Compute reconstructed image
end
```

Table 3.4 Source code for generating a phase-only hologram of a two-layer object with the noise-addition method

```
function NA()
  close all; clc; warning off
  fname='HK1';
  I=imread(strcat(fname,'.jpg'));      %Filename of intensity image
  I=im2double(I);
  [Y X]=size(I(:,:));                  %Size of source image
  N=Y*2;M=X*2;                         %Size of hologram = Two times size of
                                         %image
  A=rand(Y,X)*2*pi;                    %Random phase
  IN=I.*(cos(A)+j*sin(A));             %Add random phase to first image
  I1=zeros(N,M);                       %Declare a larger canvas to house the
                                         %source image
  I2=zeros(N,M);                       %Declare a larger canvas to house the
                                         %source image
  sy=(N-Y)/2;ey=(N+Y)/2-1; sx=(M-X)/2;ex=(M+X)/2-1;
  I1(sy:ey,sx:ex)=IN(:,:);             %Move noise-added image to center of I1
  I1(:,M/2+1:M)=0;                     %Only left half-plane of source image in
                                         %I1
  I2(sy:ey,sx:ex)=IN(:,:);             %Move noise-added image to center of I2
  I2(:,1:M/2)=0;                       %Only right half-plane of source image
                                         %in I2
  psize=6.4e-6;                        %Pixel size of hologram=6.4 µm
  lamda=520e-9;                        %Wavelength=520 nm
  z1=0.07;                             %Depth of left half-plane of image (first
                                         %layer)
  z2=0.08;                             %Depth of right half-plane of image
                                         %(second layer)
  H1=Holo_gen(I1,z1,psize,lamda);      %Hologram of first layer
  H2=Holo_gen(I2,z2,psize,lamda);      %Hologram of second layer
  H=H1+H2;                             %Hologram=sum of first and second layer
                                         %holograms
  HP=cos(angle(H))+j*sin(angle(H));    %Remove magnitude of hologram
  %Display and save phase-only hologram
  A=mat2gray(angle(HP));
  figure; imshow(A);
  title('Phase-only hologram generated with the noise-addition method');
  imwrite(A,'NA_EDGE_POH.jpg');
  %Reconstruct images of the phase-only hologram on the two depth planes
  RES1=mat2gray(abs(Holo_recon(HP,z1,psize,lamda)));
  RES_Corp1=RES1(sy:ey,sx:ex);
  RES2=mat2gray(abs(Holo_recon(HP,z2,psize,lamda)));
  RES_Corp2=RES2(sy:ey,sx:ex);
  imwrite(RES_Corp1,'RES_NA_EDGE_D1.jpg');
  imwrite(RES_Corp2,'RES_NA_EDGE_D2.jpg');
  figure,imshow(RES_Corp1);            %Display reconstructed image on first
                                         %depth plane
  title('Reconstructed image on first depth plane');
```

Table 3.4 (cont.)

```
  figure,imshow(RES_Corp2);              %Display reconstructed image on second
                                           %depth plane
  title('Reconstructed image on second depth plane');
end
function H=Holo_gen(I,z,psize,lamda)
  [N M]=size(I(:,:));
  wn=2*pi/lamda;                         %Wave number
  %Generate Fresnel zone plate
  z2=z*z;                                %Square of the depth value z
  [kx ky]=meshgrid(-M/2:M/2-1,-N/2:N/2-1);
  kx=kx*psize;                           %Physical horizontal position
  ky=ky*psize;                           %Physical vertical position
  fzp=exp(j*wn*sqrt(kx.^2+ky.^2+z2));
  %Compute hologram of source image
  A=fft2(I);                             %Fourier transform of the hologram
  B=fft2(fzp);                           %Fourier transform of the free space
                                           %impulse response
  H=fftshift(ifft2(A.*B));
end
function IR=Holo_recon(H,z,psize,lamda)
  [N M]=size(H(:,:));                    %Size of hologram
  wn=2*pi/lamda;                         %Wave number
  %Generate Fresnel zone plate
  z2=z*z;                                %Square of the depth value z
  [kx ky]=meshgrid(-M/2:M/2-1,-N/2:N/2-1);
  kx=kx*psize;                           %Physical horizontal position
  ky=ky*psize;                           %Physical vertical position
  r=sqrt(kx.^2+ky.^2+z2);
  fzp_c=exp(-j*wn*r)./r;                 %Conjugate of FZP
  A=fft2(H);                             %Fourier transform of the hologram
  B=fft2(fzp_c);                         %Fourier transform of the free space
                                           %impulse response
  IR=fftshift(ifft2(A.*B));              %Compute reconstructed image
end
```

3.4.3 Simulation of Generating a Phase-Only Hologram of a Two-Layer Object (Double-Depth Image) with the Noise-Addition Method

In this simulation, with the source code given in Table 3.4, the phase-only holograms of an object space comprising a double-depth image is generated with the noise-addition method. The double-depth image is composed of two planar images, each of which is parallel to, and located at a different distance from, the hologram. In the simulation, the main function "NA," is called to load the intensity image of the source image file.

Table 3.5 Source code for generating a patterned phase-only hologram

```
function PPOH()
  close all; clc; warning off
  fname='HK1';
  I=imread(strcat(fname,'.jpg'));     %Read intensity of double-depth image
  I=im2double(I);
  [Y X]=size(I(:,:));                 %Size of input image
  N=Y*2;M=X*2;                        %Size of hologram = Two times size of
                                        %image
  %Set size of first and second image planes=size of hologram plane
  I1=zeros(N,M);                      %First layer hosting left half-plane of
                                        %the source image
  I2=zeros(N,M);                      %Second layer hosting right half-plane
                                        %of the source image
  sy=(N-Y)/2;ey=(N+Y)/2-1; sx=(M-X)/2;ex=(M+X)/2-1;
  I1(sy:ey,sx:ex)=I(:,:);             %Move source image to center of first
                                        %layer
  I1(:,M/2+1:M)=0;                    %This layer only contains the left half-
                                        %plane
  I2(sy:ey,sx:ex)=I(:,:);             %Move source image to center of second
                                        %layer
  I2(:,1:M/2)=0;                      %This layer only contains the right
                                        %half-plane
  psize=6.4e-6;                       %Pixel size of hologram=6.4 μm
  lamda=520e-9;                       %Wavelength=520 nm
  z1=0.07;                            %Depth plane of first layer (left
                                        %section) at 0.07 m
  z2=0.08;                            %Depth plane of second layer (right
                                        %section) at 0.08 m
  BS=input('Enter block size ');
  PP=rand(BS,BS)*2*pi;                %Random phase pattern of size BSxBS
  %Generate phase mask from repetition of the phase pattern
  for row=1:N
   for col=1:M
     A(row,col)=PP(mod(row-1,BS)+1,mod(col-1,BS)+1);
   end
  end
  I1=I1.*(cos(A)+j*sin(A));           %Add random phase to image of first layer
  I2=I2.*(cos(A)+j*sin(A));           %Add random phase to image of second
                                        %layer
  H1=Holo_gen(I1,z1,psize,lamda);     %Generate object wave of image on first
                                        %layer
  H2=Holo_gen(I2,z2,psize,lamda);     %Generate object wave of image on second
                                        %layer
  H=H1+H2;                            %Hologram = sum of holograms of both
                                        %layers
  A=angle(H);
  HP=cos(A)+j*sin(A);                 %Remove magnitude of hologram
  A=mat2gray(A);
```

Table 3.5 (cont.)

```
  %Display and save patterned phase-only hologram
  figure;
  imshow(A);
  title('Phase-only hologram generated by the PPOH method');
  imwrite(A,strcat('PPOH_POH_BS',num2str(BS),'.jpg'));
  %Reconstruct image of the phase-only hologram at the first depth planes
  RES1=mat2gray(abs(Holo_recon(HP,z1,psize,lamda)));
  RES_Corp1=RES1(sy:ey,sx:ex);
  %Reconstruct image of the phase-only hologram at the second depth planes
  RES2=mat2gray(abs(Holo_recon(HP,z2,psize,lamda)));
  RES_Corp2=RES2(sy:ey,sx:ex);
  %Display reconstructed images on both depth planes
  figure,imshow(RES_Corp1);            %Display reconstructed image on first
                                         %depth plane
  title('Reconstructed image on depth plane z1=0.07 m');
  figure,imshow(RES_Corp2);            %Display reconstructed image on first
                                         %depth plane
  title('Reconstructed image on depth plane z1=0.08 m');
  %Save reconstructed images on both depth planes
  imwrite(RES_Corp1,strcat('RES_PPOH_',num2str(BS),'_D1.jpg'));
  imwrite(RES_Corp2,strcat('RES_PPOH_',num2str(BS),'_D2.jpg'));
end
function H=Holo_gen(I,z,psize,lamda)
  [N M]=size(I(:,:));
  wn=2*pi/lamda;                       %Wave number
  %Generate Fresnel zone plate
  z2=z*z;                              %Square of the depth value z
  [kx ky]=meshgrid(-M/2:M/2-1,-N/2:N/2-1);
  kx=kx*psize;                         %Physical horizontal position
  ky=ky*psize;                         %Physical vertical position
  fzp=exp(j*wn*sqrt(kx.^2+ky.^2+z2));
  %Compute hologram of source image
  A=fft2(I);                           %Fourier transform of the hologram
  B=fft2(fzp);                         %Fourier transform of the free space
                                         %impulse response
  H=fftshift(ifft2(A.*B));
end
function IR=Holo_recon(H,z,psize,
  lamda)
  [N M]=size(H(:,:));                  %Size of hologram
  wn=2*pi/lamda;                       %Wave number
  %Generate Fresnel zone plate
  z2=z*z;                              %Square of the depth value z
  [kx ky]=meshgrid(-M/2:M/2-1,-N/2:N/2-1);
  kx=kx*psize;                         %Physical horizontal position
  ky=ky*psize;                         %Physical vertical position
  r=sqrt(kx.^2+ky.^2+z2);
  fzp_c=exp(-j*wn*r)./r;               %Conjugate of FZP
```

Table 3.5 (cont.)

```
  A=fft2(H);                     %Fourier transform of the hologram
  B=fft2(fzp_c);                 %Fourier transform of the free space
                                   %impulse response
  IR=fftshift(ifft2(A.*B));      %Compute reconstructed image
end
```

The size of the hologram is set to twice that of the image. Next, the intensity of the source image is added with random phase noise, and the double-depth, noise-added image is converted into a hologram with the function "Holo_gen." The phase component of the hologram is retained as the phase-only hologram, and reconstructed images at the two focused planes is obtained with the function "Holo_recon." Subsequently, the reconstructed images are displayed and saved.

3.4.4 Simulation of Generating a Phase-Only Hologram of a Two-Layer Object with the PPOH Method

In this simulation, the phase-only hologram of the double-depth image "HK1" is generated with the PPOH method. The source code is shown in Table 3.5. The double-depth image is composed of two planar images, each of which is parallel to, and located at a different distance from, the hologram. In the simulation, the main function "PPOH," is called to load the intensity image from the source image file. The size of the hologram is set to two times that of the image. Next, the intensity of the double-depth image is added with the repetitive phase pattern having a period (block size) τ that is input by the user. The function "Holo_gen" is then applied to generate a hologram for the patterned phase-added double-depth image. Subsequently, the phase component of the hologram is retained as the phase-only hologram, and the reconstructed images at the two focused planes are obtained with the function "Holo_recon." The reconstructed images on the pair of focused planes are displayed and saved.

3.5 Summary

A phase-only Fresnel hologram has two major advantages over complex-valued or amplitude-only holograms. First, it can be displayed with a single phase-only SLM, leading to simplification on the holographic display system. Second, due to the high optical efficiency of the phase-only hologram, the reconstructed image is brighter than that of an amplitude-only hologram. On the downside, the fidelity of the reconstructed image is degraded as a result of discarding the magnitude component of the hologram. In this chapter, a number of methods, each with pros and cons, for generating phase-only holograms are described. These methods can be divided into two types, the iterative and the non-iterative approaches. Iterative methods include the IFTA and its variants, which find their origin in the classical Gerchberg–Saxton algorithm. Reconstructed images of a phase-only hologram obtained with IFTA are generally good in quality, but the computation time is rather lengthy. Another iterative method, based on DBS, can be applied in generating BPOHs. Non-iterative methods are based on modifying the source image in certain ways prior to the generation of its hologram. These include noise addition, patterned phase addition, and downsampling. The modification is similar to overlaying a diffuser onto the image, so that the magnitude of the diffracted waves on the hologram is close to homogeneous. The phase component alone, therefore, can be taken to represent the hologram.

Exercises

E.3.1 In Section 3.2.1, one, three, and five rounds of IFTA are applied to generate the phase-only holograms of the source image in Figure 3.4(a), and the results are shown in Figure 3.4(b–d). What will be the result if more iterations are used in the IFTA?

E.3.2 Based on the MATLAB codes provided for the mixed-region amplitude freedom (MRAF) method, evaluate the effect of the proportion between the signal region and the noise regions on the reconstructed image.

E.3.3 Implement a MATLAB program to generate a phase-only hologram of a multiple-depth image based on the MRAF method.

E.3.4 The quality of the reconstructed image of a hologram generated with the noise-addition method is affected by the random phase noise that is added to the source image. In your opinion, will it be possible to determine random phase noise that can improve the reconstructed image? If affirmative, implement the method based on the source codes provided for the noise-addition method.

E.3.5 Perform a literature reading on the OSPR method [12,13] and write a simulation program based on MATLAB for applying the method in generating a BPOH.

E.3.6 In generating a patterned phase-only hologram, different phase patterns result in different local properties of the reconstructed image. Suggest one or a few methods to determine a set of optimal phase patterns.

E.3.7 Based on the MATLAB program in Section 3.4.4, explore the effects of the following items on the reconstructed image of the patterned phase-only hologram:

a. block size;
b. axial distance between the source image and the hologram;
c. hologram size.

E.3.8 Write a MATLAB program for generating an SPOH, and explore the effects of the following items on its reconstructed image:

a. block size;

b. axial distance between the source image and the hologram;
c. sampling patterns other than the grid-cross lattice;
d. hologram size.

References

[1] C. Stolz, L. Bigue, and P. Ambs, "Implementation of high-resolution diffractive optical elements on coupled phase and amplitude spatial light modulators," *Appl. Opt.*, vol. 40, no. 35, pp. 6415–6424 (2001).

[2] R. Tudela, I. Labastida, E. Martín-Badosa, S. Vallmitjana, I. Juvells, and A. Carnicer, "A simple method for displaying Fresnel holograms on liquid crystal panels," *Opt. Commun.*, vol. 214, pp. 107–114 (2002).

[3] L. Zhu and J. Wang, "Arbitrary manipulation of spatial amplitude and phase using phase-only spatial light modulators," *Sci. Rep.*, vol. 4, art. no. 7441 (2014).

[4] A. Siemion, M. Sypek, J. Suszek, M. Makowski, A. Siemion, A. Kolodziejczyk, and Z. Jaroszewicz, "Diffuserless holographic projection working on twin spatial light modulators," *Opt. Lett.*, vol. 37, pp. 5064–5066 (2012).

[5] M. Makowski, A. Siemion, I. Ducin, K. Kakarenko, M. Sypek, A. Siemion, J. Suszek, D. Wojnowski, Z. Jaroszewicz, and A. Kolodziejczyk, "Complex light modulation for lensless image projection," *Chin. Opt. Lett.*, vol. 9, art. no. 120008 (2011).

[6] J.-P. Liu, W.-Y. Hsieh, T.-C. Poon, and P. W. M. Tsang, "Complex Fresnel hologram display using a single SLM," *Appl. Opt.*, vol. 50, no. 34, pp. H128–H135 (2011).

[7] H. Song, G. Sung, S. Choi, K. Won, H. Lee, and H. Kim, "Optimal synthesis of double-phase computer generated holograms using a phase-only spatial light modulator with grating filter," *Opt. Exp.*, vol. 20, pp. 29844–29853 (2012).

[8] R. W. Gerchberg and W. O. Saxton, "A practical algorithm for the determination of the phase from image and diffraction plane pictures," *Optik (Stuttg.)*, vol. 35, pp. 237–246 (1972).

[9] J. Yeom, J. Hong, J. Jung, K. Hong, J. Park, and B. Lee, "Phase-only hologram generation based on integral imaging and its enhancement in depth resolution," *Chin. Opt. Lett.*, vol. 9, art no. 120009 (2011).

[10] M. Pasienski and B. DeMarco, "A high-accuracy algorithm for designing arbitrary holographic atom traps," *Opt. Express*, vol. 16, pp. 2176–2190 (2008).

[11] M. Makowski, Ma. Sypek, A. Kolodziejczyk, and G. Mikula "Three-plane phase-only computer hologram generated with iterative Fresnel algorithm," *Opt. Eng.*, vol. 44, no. 12, art no. 125805 (2005).

[12] P.W.M. Tsang, Y.T. Chow, and T.-C. Poon, "Generation of edge-preserved noise-added phase-only hologram," *Chin. Opt. Lett.*, vol. 14, art no. 100901 (2016).

[13] E. Buckley, "Holographic laser projection technology." Proc. SID Symp., 1074–1078 (2008).

[14] E. Buckley, "Real-time error diffusion for signal-to-noise ratio improvement in a holographic projection system," *J. Disp. Tech.*, vol. 7, no. 2, pp. 70–76 (2011).

[15] P.W.M. Tsang, Y.-T. Chow, and T.-C. Poon, "Generation of patterned-phase-only holograms (PPOHs)," *Opt. Express*, vol. 25, pp. 9088–9093 (2017).

[16] P.W.M. Tsang, Y.-T. Chow, and T.-C. Poon, "Generation of phase-only Fresnel hologram based on down-sampling," *Opt. Express*, vol. 22, pp. 25208–25214 (2014).

[17] P.W.M. Tsang, Y.-T. Chow, and T.-C. Poon, "Enhancement on the generation of sampled phase-only holograms," *Chin. Opt. Lett.*, vol. 13, art no. 060901 (2015).

[18] P.W.M. Tsang, Y.-T. Chow, and T.-C. Poon, "Generation of complementary sampled phase-only holograms," *Opt. Express*, vol. 24, pp. 23390–23395 (2016).

[19] T. Leportier and M.C. Park, "Generation of binary holograms for deep scenes captured with a camera and a depth sensor," *Opt. Eng.*, vol. 56, art no. 013107 (2017).

[20] J.-P. Liu, C. Yu, and P.W.M. Tsang, "Enhanced direct binary search algorithm for binary computer-generated Fresnel holograms," *Appl. Opt.*, vol. 58, pp. 3735–3741 (2019).

4 Conversion of Complex-Valued Holograms to Phase-Only Holograms

4.1 Introduction

In previous chapters, different methods for generating a phase-only hologram for a given 3-D object, such as the iterative Fresnel transform algorithm (IFTA), down-sampling, and the patterned phase method, have been presented. While all these methods are effective in generating phase-only holograms, they all share the same prerequisite that a source image has to be available in the first place. This is not a problem in computer-generated holography (CGH), in which case the source images are provided in the form of numerical data, such as point cloud files or computer graphic models. However, in some applications such as remote sensing and microscopic holography, a hologram is captured directly from a real object based on acquisition methods like optical scanning holography and phase-shifting holography. For these cases, the source image is not available, and the methods that have been gone through previously are not applicable for deriving the phase-only hologram.

Intuitively, it could be possible to reconstruct the 3-D source image from a complex-valued hologram, with which a phase-only hologram can be generated with IFTA or other methods. However, there is a lack of an effective method that can reconstruct an arbitrary 3-D image from its hologram. In most of the methods developed to date, a hologram is reconstructed onto a series of regularly spaced depth planes. Subsequently, some sort of sharpness or focus index is measured for each region on the stack of reconstructed images, and the depth plane with the sharpest measurement among all its peers will be taken to be the focused plane of that particular region. Although such an approach is logical and quite sound, existing hologram reconstruction methods such as those discussed in [1–4] are computation intensive, and so only of use for reconstructing simple geometries. For more complicated object shapes, a region is likely to contain contents with a wider range of depths. When one area within a region is in focus, the remaining part may become blurred and the out-of-focus content may blend into the neighboring focused region(s). It is therefore not possible to use the sharpness index to locate the focused plane of each region unless the object space is thin or comprises of widely separated items.

Alternatively, it is possible to convert a complex-valued hologram into a phase-only hologram. There are two major difficulties with this approach. First, the process is lossy and irreversible, as the magnitude component of the hologram is removed. The amount of distortion incurred in the conversion is unpredictable. Second, as there is no information on the source image that is represented by the hologram, iterative techniques (for example, the Gerchberg–Saxton algorithm) cannot be applied to minimize the conversion error. This chapter describes six methods for converting a complex-valued hologram into a phase-only hologram: complex amplitude modulation (CAM); double-phase decomposition; uni-directional error diffusion (UERD); bi-directional error diffusion (BERD); localized error diffusion (LERD); and direct binary search (DBS). This last method is used for obtaining a binary phase-only hologram from the complex-valued hologram.

4.2 Complex Amplitude Modulation

Complex amplitude modulation is a method of encoding both the amplitude and phase components of a complex-valued function into a phase-only function. The idea of applying this concept in digital holography could be dated back to 1999, when J.A. Davis and his team succeeded in implementing phase-only optical filters for optical recognition and image processing [5]. Recently, the method has been applied to convert a complex-valued hologram into a phase-only hologram [6,7]. Upon illumination with a reference beam and filtering of the diffracted waves, the source image recorded in the phase-only hologram will be reconstructed as a visible image.

Here, a complex-valued hologram $H(m,n) = |H(m,n)|\exp^{i\theta(m,n)}$ is to be converted into a phase-only hologram $H_p(m,n)$. After the hologram is generated, it will be illuminated by an inclined plane wave $R(m,n) = |R(m,n)|\exp^{i\theta_R(m,n)}$, making an angle of incidence θ_R along the vertical or horizontal direction for reconstructing the image recorded in the hologram.

Converting the hologram into a phase-only hologram with CAM is straightforward. The phase-only hologram is obtained simply by integrating the hologram $H(m,n)$ and the off-axis plane wave into a phase function:

$$H_P(m,n) = c\exp^{\{i\beta|H(m,n)|\cos[\theta(m,n)-\theta_R(m,n)]\}}, \tag{4.1}$$

where c and β are constant terms.

To view the reconstructed image, the phase-only hologram $H_P(m,n)$ is illuminated by the inclined reference plane wave $R(m,n)$ that is making an angle of incidence θ_R along the vertical or horizontal direction. The diffracted wave on the reconstructed image plane is

$$D_P(m,n) = c\exp^{i\theta_R(m,n)}\exp^{\{i\beta|H(m,n)|\cos[\theta(m,n)-\theta_R(m,n)]\}}. \tag{4.2}$$

From Eq. (4.2) it can be seen that although the expression should contain the signal corresponding to the hologram $H(m,n)$, the diffracted wave $D_P(m,n)$ also includes components that are different from the hologram. The question is: How can the hologram signal $H(m,n)$ be separated from the rest of the terms in $D_P(m,n)$. It is shown in [6] that this is achieved by decomposing the expression in Eq. (4.2) into an infinite series based on the following property of the Bessel function:

$$\exp^{is\cos\psi} = \sum_{k=-\infty}^{\infty} i^k J_k(s)\exp^{ik\psi}, \tag{4.3}$$

where $J_k(s)$ is the kth order of the Bessel function of the first kind.

Applying Eq. (4.3) to Eq. (4.2), the diffraction wave $D_P(m,n)$ is the sum of an infinite series of functions:

$$D_P(m,n) = c\sum_{-\infty}^{\infty} J_k[\beta|H(m,n)|]i^k\exp^{\{-i[k\theta(m,n)-(k+1)\theta_R(m,n)]\}}. \tag{4.4}$$

Eq. (4.4) is equivalent to the superposition of an infinite number of off-axis holograms; the reconstructed image of each one projects at a unique angle, which is an integer multiple of the off-axis illumination angle θ_R of the reference plane wave. For simplicity, the constant term c can be taken as unity.

In Eq. (4.4), the component corresponding to $k = -1$,

$$H_{CAM}(m,n) = D_P(m,n)|_{k=-1} = -J_{-1}[\beta|H(m,n)|]\exp^{i\theta_R(m,n)}. \tag{4.5}$$

The reconstructed images of higher-order diffraction with $|k| > 1$ should be non-overlapping with the one corresponding to $k = -1$ if the angle of incidence θ_R of the reference wave is sufficiently large. Based on this assumption, an optical mask can be applied to extract $H_{CAM}(m,n)$ from the rest of the unwanted components. From Eq. (4.5), the -1 order of diffraction (denoted by $H_{CAM}(m,n)$) is related to the complex amplitude distribution of the hologram $H(m,n)$. If $J_{-1}[|H(m,n)|]$ is similar to $H(m,n)$, the source image can be reconstructed to a good approximation from this component. The reconstructed image of $H_{CAM}(m,n)$ will carry a certain amount of distortion, as the hologram has been changed to the function $J_{-1}[|H(m,n)|]$, and the reconstructed image is overlaid with the zero-order ($k = 0$).

The first type of distortion, resulting from the modification of the hologram by $J_{-1}[|H(m,n)|]$, can be alleviated by selecting a value for the constant term β that restricts the dynamic range of $|H(m,n)|$ to the linear region of the Bessel function. To illustrate this point, a plot of the function $J_{-1}[x]$ for $x = [0,5]$ is shown in Figure 4.1. From the diagram, the Bessel function is close to a linear function for x within the range [0,1.5].

The distortion caused by the overlaying of the zero-order component on the reconstructed image can be removed with an optical filter formed by a pair of lens arranged in a 4f filtering configuration.

Numerical simulation of the CAM method is conducted using the MATLAB code provided in Section 4.8.1. A complex-valued Fresnel hologram of the source image "Apple" in Figure 4.2(a) is first generated. Next, the phase-only hologram corresponding to the -1 order of diffraction signal $H_{CAM}(m,n)$ is extracted with Eq. (4.5). A

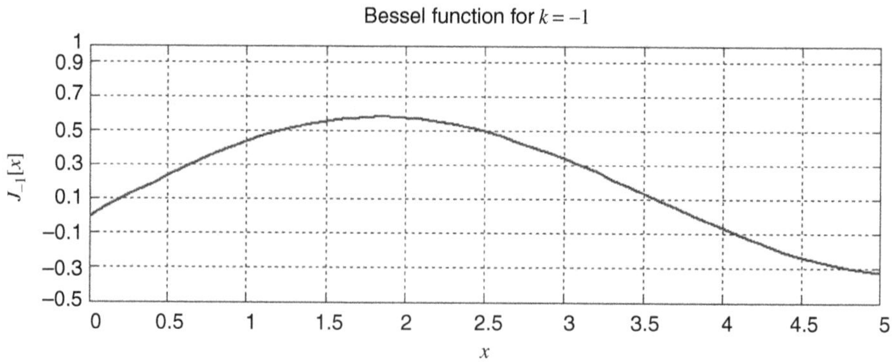

Figure 4.1 Plot of Bessel function $J_{-1}[x]$ for $x = [0,5]$.

reconstructed image of the phase-only hologram on the focused plane is thus obtained. The reconstructed images corresponding to $\beta = 3.0$, 2.0, and 0.50 are shown in Figures 4.2(b–d). In all these images, the dynamic range of brightness has been normalized to the range $\{0, 255\}$, with 0 and 255 representing the lowest and the highest intensities, respectively. The attenuation of the brightness due to β, therefore, is not shown in the figures. The fidelities of the three reconstructed images, as compared with the original one in Figure 4.2(a), are 14.45 dB, 20.89 dB, and 30 dB for $\beta = 3$, $\beta = 2$, and $\beta = 0.5$. In general, the larger the peak-signal-to-noise-ratio (PSNR), the higher is the fidelity of the image. By observation, the quality of the reconstructed image is increased with decreasing values of β, which is also reflected by the increase in PSNR. However, from Eq. (4.5) and Figure 4.1, the smaller the value of β, the weaker the magnitude of the hologram $H_{CAM}(m, n)$, and the dimmer will be its reconstructed image. Hence it is necessary to find a value of β that results in a reconstructed image with higher fidelity and sufficient brightness to create a favorable visual quality.

Another similar simulation is applied to convert the complex-valued hologram of the double-depth image "HK1" in Figure 1.12 into a phase-only hologram with the CAM method. The source image is divided into left and right sections located at axial distances of $z_1 = 0.07$ m and $z_2 = 0.08$ m from the hologram. A phase-only hologram is generated with $\beta = 0.5$, and the reconstructed images at the two focused planes are shown in Figure 4.3. From these two images, each side of the source image is reconstructed as a sharp image on its focused plane, with the other side being a blurred, defocused image. This simulation result demonstrates that the CAM method is capable of generating a phase-only hologram that can represent both intensity and depth dimension.

4.3 Double-Phase Macro-Pixel Hologram

The double-phase macro-pixel hologram is based on the principle that if a complex number is bounded within the unit circle, it can be represented as the sum of two phase-only quantities. As a result, any complex quantity can be converted into a pure phase

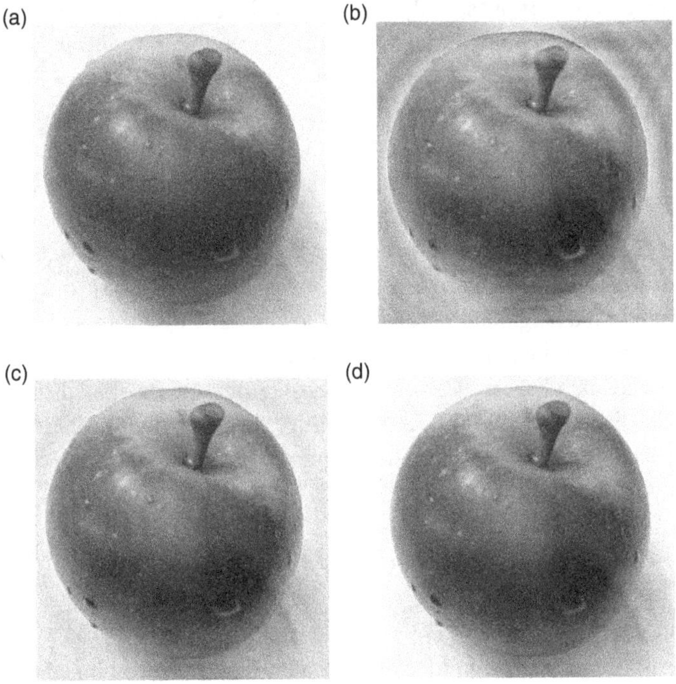

Figure 4.2 (a) Source image "Apple." (b–d) Reconstructed image from $H_{CAM}(m,n)$ with $\beta = 3, 2, 0.5$, respectively.

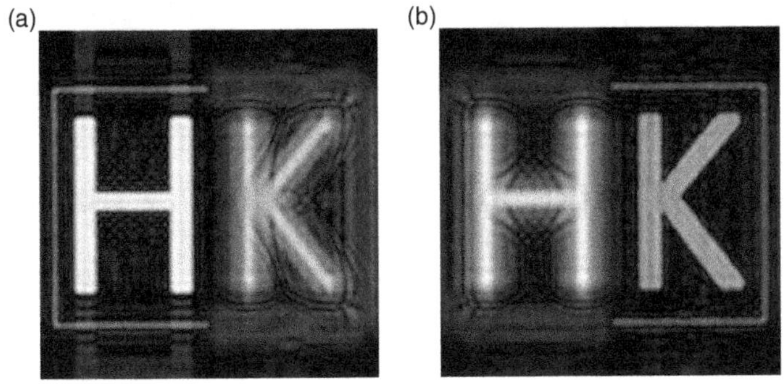

Figure 4.3 Reconstructed images of the phase-only hologram, obtained with the CAM method with $\beta = 0.5$. (a) On focused plane at 0.07 m; (b) on focused plane at 0.08 m.

representation. Mathematically, given a complex number $A\exp^{i\theta}$ with A within the range [0,1], the sum of two phase terms is expressed as

$$A\exp^{i\theta} = 0.5\left[\exp^{i\theta_1} + \exp^{i\theta_2}\right], \tag{4.6}$$

where $A = \cos(\alpha)$, $\theta_1 = \theta + \alpha$, and $\theta_2 = \theta - \alpha$. The simple mathematical relation in Eq. (4.6) suggests that this principle can be applied to convert a complex-valued hologram into a pair of phase-only images, which could be integrated in certain ways to form a double-phase-only hologram (DPH). From Eq. (4.6), it can be seen that as A is positive, the angle $\alpha = \cos^{-1}(A)$ is confined to the right half of the Euclidean plane.

The generation of a double-phase hologram was achieved by the pioneering attempt of C. Hsueh and A. Sawchuk [8] in the mid-1970s. In their investigation, a Fourier hologram was decomposed into a pair of phase components based on Eq. (4.6). Next, each Fourier cell (which can be interpreted as a hologram pixel) was split into two sub-cells, each encoding one of the phase terms. In the simplest case, a Fourier cell is evenly separated into two partitions, as shown in Figure 4.4(a). In each partition, a window of half the length of the cell is shifted vertically by an amount (τ_1 or τ_2), which is proportionally to the phase quantity it represents.

With the availability of electronic accessible phase-only spatial light modulators (SLMs), the display of DPH has been simplified. In the work of J.M. Florence and R.D. Juday [9], a complex-valued Fresnel hologram $H(m,n)$ was decomposed into a pair of phase-only images, $\theta_1(m,n)$ and $\theta_2(m,n)$:

$$H(m,n) = |H(m,n)|\exp^{i\theta(m,n)} = 0.5\left[\exp^{i\theta_1(m,n)} + \exp^{i\theta_2(m,n)}\right]. \tag{4.7}$$

Although both exponential terms on the right-hand side of Eq. (4.7) are phase-only quantities with constant magnitude of unity, they have to be combined to the complex-valued hologram. To address this issue, the two phase images are combined into a double-phase hologram $\theta_M(m,n)$ through spatial interleaving. A straightforward way to interleave the pair of phase images is to copying the odd and even rows of $\theta_1(m,n)$ and $\theta_2(m,n)$ to the odd and even rows of $\theta_M(m,n)$, respectively. The adjacent pair of pixels in $\theta_M(m,n)$ forms a 1×2 macro-pixel, with each of its element (referred to as a sub-pixel) contributing from either $\theta_1(m,n)$ or $\theta_2(m,n)$, as shown in Figure 4.4(b). In another words, the pair of phase images are each downsampled by two times and placed at non-overlapping areas in $\theta_M(m,n)$.

When the double-phase hologram is illuminated by a coherent beam, the diffracted waves corresponding to $\theta_1(m,n)$ and $\theta_2(m,n)$ will be generated and merged on the reconstruction plane. However, due to the downsampling of these two signals, their frequency spectra will be added with aliasing images. In order to attenuate the aliasing images, and to merge $\theta_1(m,n)$ and $\theta_2(m,n)$, certain optical filtering or processing has to be applied so that the reconstructed image will be similar to that of the original hologram $H(m,n)$. As pointed out by V. Arrizón and D. Sanchez-de-la-Llave [10], the region in the reconstruction plane having high fidelity (measured in terms of signal-to-noise ratio) is rather small along the direction of the macro-pixel. To enhance the fidelity, the symmetrical 2×2 distribution of sub-pixels in Figure 4.4(c) is often adopted in forming a macro-pixel.

In the 2×2 macro-pixel arrangement, the pair of phase components $\theta_1(m,n)$ and $\theta_2(m,n)$ are downsampled by the orthogonal uniform sampling lattices $L_1(m,n)$ and $L_2(m,n)$, as shown in Figure 4.5(a) and 4.5(b). Each lattice is in the form of a checkerboard pattern, with the sampled and non-sampled points assigned values of

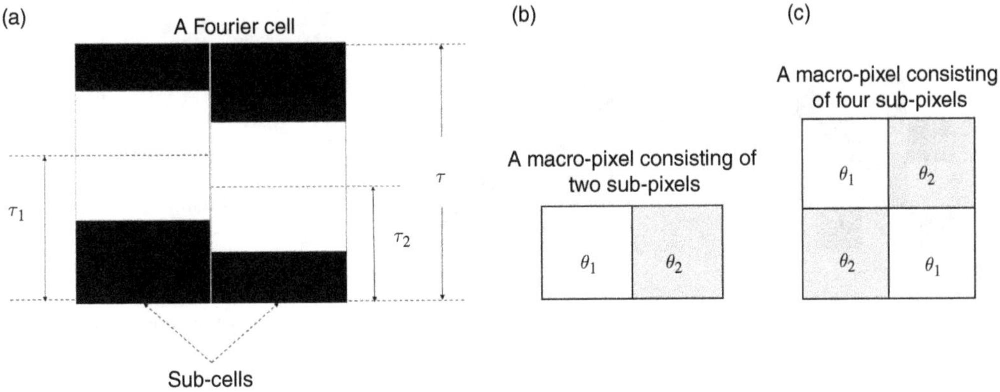

Figure 4.4 (a) A Fourier cell with two vertical shifted sub-cells for representing two phase quantities. (b) Interleaving a pair of phase components with the 1 × 2 macro-pixel topology. (c) Interleaving a pair of phase components with the 2 × 2 macro-pixel topology.

1 (white) and 0 (black), respectively. The result of multiplexing $\theta_1(m,n)$ and $\theta_2(m,n)$ with the two sampling lattices is shown in Figure 4.5(c):

$$\theta_M(m,n) = L_1(m,n)\theta_1(m,n) + L_2(m,n)\theta_2(m,n). \tag{4.8}$$

The Fourier transform of $\theta_M(m,n)$ is given by

$$\tilde{\theta}_M(\omega_m,\omega_n) = \tilde{L}_1(\omega_m,\omega_n) * \tilde{\theta}_1(\omega_m,\omega_n) + \tilde{L}_2(\omega_m,\omega_n) * \tilde{\theta}_2(\omega_m,\omega_n), \tag{4.9}$$

where $\tilde{B}(\omega_m,\omega_n)$ denotes the Fourier transform of an arbitrary function $B(m,n)$, and $*$ is the convolution operator. Eq. (4.9) is therefore rewritten as

$$\begin{aligned}\tilde{\theta}_M(\omega_m,\omega_n) &= \left[\tilde{\theta}_1(\omega_m,\omega_n) + \tilde{\theta}_2(\omega_m,\omega_n)\right] + \tilde{A}_S(\omega_m,\omega_n) \\ &= \tilde{H}(\omega_m,\omega_n) + \tilde{A}_S(\omega_m,\omega_n).\end{aligned} \tag{4.10}$$

The first term on the right-hand side of Eq. (4.10) is the Fourier transform of the original hologram $H(m,n)$, while the rest of the terms are aliasing components comprising of shifted replicas of the baseband signal. If the sampling interval is small enough, the aliasing components are widely spaced so that they could be suppressed to a large extent with a low-pass filter. For example, an optical filter realized with a standard 4f configuration and an iris around the center of the Fourier plane can be employed to extract a low-pass version $H_L(m,n)$ of the original complex hologram signal.

Numerical simulation on the double-phase macroblock method in converting a complex-valued hologram into a phase-only hologram is conducted with the MATLAB code in Section 4.8.2. First, a complex-valued hologram is generated for the double-depth image "HK1" in Figure 1.12. Next, the complex-valued hologram is converted into a pair of phase-only holograms with the double-phase macroblock method, based on the 1 × 2 and the 2 × 2 macroblock topologies. Subsequently,

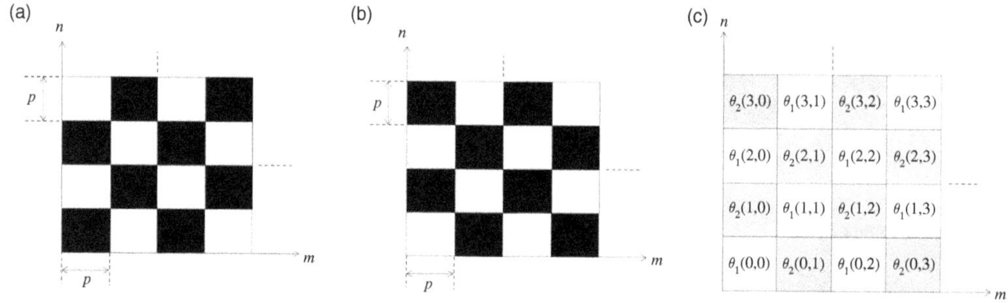

Figure 4.5 (a) Sampling lattice $L_1(m,n)$. (b) Sampling lattice $L_2(m,n)$. (c) Multiplexing $\theta_1(m,n)$ and $\theta_2(m,n)$ with $L_1(m,n)$ and $L_2(m,n)$.

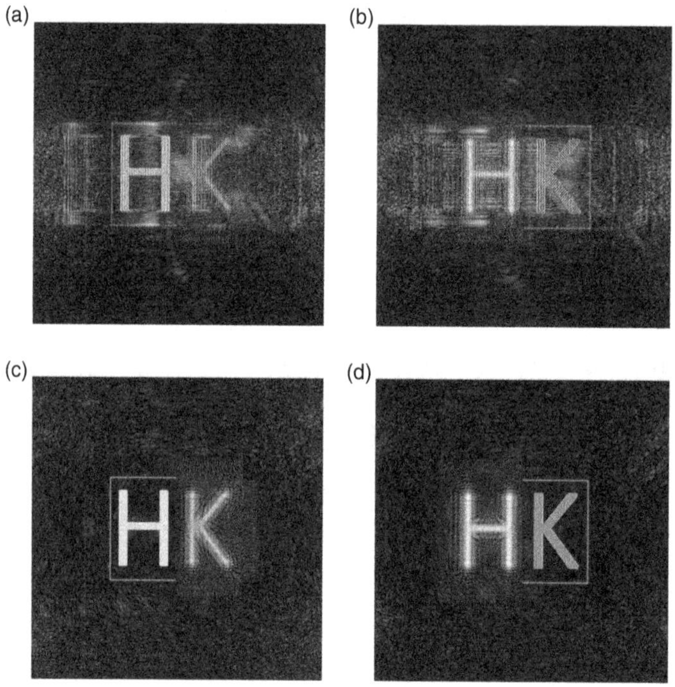

Figure 4.6 Reconstructed images on the two focused planes from the low-pass double-phase hologram: (a) 1 × 2 macro-pixel topology, focused plane at 0.07 m; (b) 1 × 2 macro-pixel topology, focused plane at 0.08 m; (c) 2 × 2 macro-pixel topology, focused plane at 0.07 m; (d) 2 × 2 macro-pixel topology, focused plane at 0.08 m.

each phase-only hologram is low-pass filtered with the bandwidth limited to one-quarter of its original value, so that aliasing errors at the high frequency range are suppressed. The reconstructed images at the two focused planes of the phase holograms generated with the 1 × 2 topology and the 2 × 2 macroblock topologies are shown in Figures 4.6(a, b) and 4.6(c, d), respectively.

First we inspect the reconstructed images in Figure 4.6(a, b) of the double-phase hologram that has been interleaved with the 1×2 macro-pixel topology. Each side of the image is reconstructed clearly from the hologram on its corresponding focused plane. However, the images are quite heavily contaminated with noise. In the reconstructed images in Figure 4.6(c, d), which are derived from the double-phase hologram interleaved with the 2×2 topology, the noise is not prominent. These results show that the 2×2 interleaving method is preferable in the generation of the double-phase hologram.

4.4 Uni-directional Error Diffusion

Although the CAM method and the double-phase hologram are both effective in converting a complex-valued Fresnel hologram into a phase-only hologram, some optical accessories (such as an optical mask and optical filter) are required to suppress the unwanted signals in the display process. This will lead to greater complexity, size, and cost of the display systems, imposing limitations on practical applications. Recently, P. Tsang and T.-C. Poon have proposed employing error diffusion for converting a Fresnel hologram into a phase-only hologram. A phase-only hologram derived with this method is referred to as a uni-directional error diffusion (UERD) hologram [11]. There are two major advantages of the UERD hologram. First, the computation time that is required to convert a complex-valued hologram to a phase-only hologram is short. Second, the phase-only hologram obtained with the UERD method can be optically reconstructed simply by illuminating the hologram with a coherent beam without additional optical processing.

Error diffusion is a technique that was originally developed for quantizing pixels of a continuous tone image with a small number of quantization levels. It has been widely applied for many decades in facsimile transmission and printing, where most of the equipment can only print binary images comprising pixels of either black or white. In general, the visual quality of a binary image that is converted with error diffusion from a continuous tone image, is preserved favorably. There are many variations in the error diffusion algorithm, the most notable one being perhaps the Floyd–Steinberg algorithm [12]. In this approach, the pixels in an image are processed sequentially in a row-by-row, left-to-right manner. Each visited pixel will be changed to a white pixel if its intensity exceeds a certain threshold, and changed to a black pixel otherwise. The error resulting from the bi-level quantization is distributed, with predefined proportions, to the four adjacent neighbors of the pixel. As the direction of processing the pixels is identical in each row, this is referred to as uni-directional error diffusion.

In [11], the principles of the Floyd–Steinberg error diffusion technique is applied to obtain the phase-only hologram. Given a complex-valued Fresnel $H(m,n)$ with the dynamic range of the magnitude normalized to $[0,1]$, each pixel is scanned sequentially in a row-by-row, left-to-right manner. The magnitude of the pixel under evaluation is forced to unity, resulting in a phase-only component, and the error is computed and

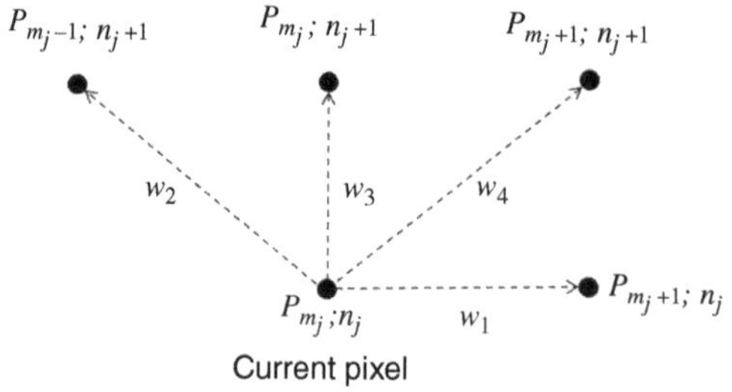

Figure 4.7 Diffusion of error in the current hologram pixel to its four immediate neighbors.

distributed to its four immediate neighbors. The process is repeated until all the pixels have been converted into a phase-only hologram $H_P(m,n)$.

Suppose $P_{m_j;n_j}$ denotes the current pixel that is being visited at (m_j, n_j); the error that results from forcing its magnitude to unity will be

$$E(m_j, n_j) = H(m_j, n_j) - H_P(m_j, n_j). \tag{4.11}$$

The error $E(m_j, n_j)$ is then added, with preset weighting factors, to the four adjacent pixels that have not been visited yet, as shown in Figure 4.7.

Mathematically, the values of these four pixels will be updated according to the following equations:

$$H(m_j+1, n_j) \leftarrow H(m_j+1, n_j) + w_1 E(m_j, n_j), \tag{4.12}$$

$$H(m_j-1, n_j+1) \leftarrow H(m_j-1, n_j+1) + w_2 E(m_j, n_j), \tag{4.13}$$

$$H(m_j, n_j+1) \leftarrow H(m_j, n_j+1) + w_3 E(m_j, n_j), \tag{4.14}$$

$$H(m_j+1, n_j+1) \leftarrow H(m_j+1, n_j+1) + w_4 E(m_j, n_j), \tag{4.15}$$

where $B \leftarrow A$ denotes replacing the value of B with the value of A. In [11], the weighting factors w_1 to w_3 follow the values adopted in the original Floyd–Steinberg algorithm, with $w_1 = \frac{7}{16}$, $w_2 = \frac{3}{16}$, $w_3 = \frac{5}{16}$, and $w_4 = \frac{1}{16}$. Optimizing the weighting factors for further enhancing the fidelity of the reconstructed image of the phase-only hologram was attempted in [13].

It can be seen from the above equations that error diffusion is a recursive process whereby pixels are processed sequentially, and non-visited pixels are modified by the errors imparted from those that have been processed. Compared with the CAM and the double-phase method, error diffusion requires a greater amount of computation, and its recursive nature prohibits the use of parallel computation. However, an error diffusion hologram can be displayed directly by illuminating it with a coherent beam, and no other optical accessories such as optical filtering and masking are required.

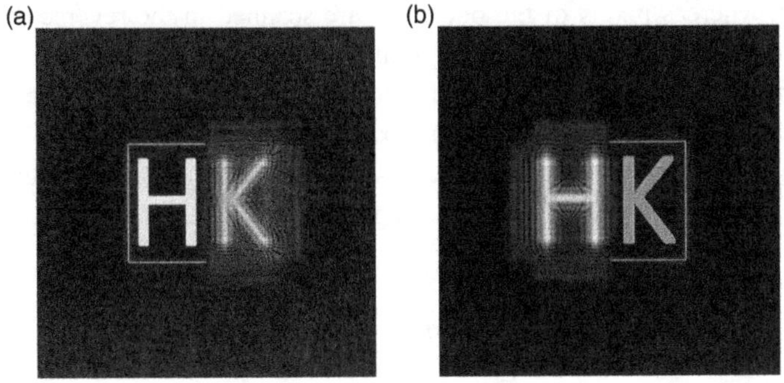

Figure 4.8 Reconstructed image from the UERD hologram: (a) on focused plane at 0.07 m; (b) on focused plane at 0.08 m.

Numerical simulation of the UERD method is conducted with the MATLAB codes in Section 4.8.3. In the simulation, a complex-valued Fresnel hologram $H(m,n)$ of the double-depth source image "HK1" is first generated. Next, Eqs. (4.12)–(4.15) are applied to convert $H(m,n)$ into a phase-only hologram $H_P(m,n)$. Finally, the left and the right sides of the phase-only hologram are each reconstructed with Fresnel diffraction on its corresponding focused plane. The reconstructed images are displayed and shown in Figure 4.8. It can be observed that apart from some noise contamination, the reconstructed image is similar to the original one. The CPU time taken in the error diffusion, based on a typical PC, is around 36 s.

4.5 Bi-directional Error Diffusion

In the conversion of a complex-valued hologram to a phase-only hologram with UERD, the error that is caused by discarding the magnitude component of each hologram pixel is diffused to the pixel at its right and the three pixels underneath. The accumulated error $E_{ACC}(m_j,n_j)$ received by a pixel at location (m_j,n_j) can be calculated with a recursive process:

$$\begin{aligned} E_{ACC}(m_j,n_j) &= w_1E(m_j-1,n_j) + w_4E(m_j-1,n_j-1) \\ &\quad + w_3E(m_j,n_j-1) + w_2E(m_j+1,n_j-1). \end{aligned} \tag{4.16}$$

Eq. (4.16) infers that the errors are propagating, and to a certain extent accumulating along the down-diagonal direction. This has the effect of errors being correlated along the line of propagation and becoming less separable from the signal. The bi-directional error diffusion (BERD) [11] is a method to break the directional behavior of error propagation, randomizing the probability distribution of the errors to make it less correlated with the signal.

Similar to UERD, pixels in a complex-valued hologram are visited in a row-by-row manner. For the even rows (i.e., rows 0, 2, 4, etc.), the pixels are scanned from left to the right, whereas pixels of the odd rows are scanned in the reverse direction. The magnitude of each visited pixel is set to unity, and the resulting error is diffused to its four immediate neighbors. The distribution of error from a pixel to its neighbors is identical to UERD for the even rows of pixels (i.e., the same as Figure 4.7). For the odd rows, the direction of scanning is reversed, and the distribution of error from the current pixel $P_{m_j;n_j}$ to the four neighboring pixels is given in following equations:

$$H(m_j - 1, n_j) \leftarrow H(m_j - 1, n_j) + w_1 E(m_j, n_j), \tag{4.17}$$

$$H(m_j + 1, n_j + 1) \leftarrow H(m_j + 1, n_j + 1) + w_2 E(m_j, n_j), \tag{4.18}$$

$$H(m_j, n_j + 1) \leftarrow H(m_j, n_j + 1) + w_3 E(m_j, n_j), \tag{4.19}$$

$$H(m_j - 1, n_j + 1) \leftarrow H(m_j - 1, n_j + 1) + w_4 E(m_j, n_j). \tag{4.20}$$

Numerical simulation on the BERD method is conducted with the MATLAB codes given in Section 4.8.4. In the simulation, a complex-valued Fresnel hologram $H(m,n)$ of the double-depth source image "HK1" is first generated. Next, Eqs. (4.17)–(4.20) are applied to convert $H(m,n)$ into a phase-only hologram $H_P(m,n)$. Finally, the left and the right sides of the phase-only hologram are each reconstructed with Fresnel diffraction on its corresponding focused plane. The reconstructed images are displayed as shown in Figure 4.9. They show that the numerical reconstructed images are similar to the original one, and the noise contamination is less than that of the UERD hologram.

Despite the high quality of the reconstructed image of an inline BERD or a UERD hologram, the optical reconstructed image is rather poor. One of the reasons is that the error diffusion process requires very precise compensation of the error in one pixel by its neighbors. While this can be resolved by numerical computation, it is not implementable in practice as the SLMs that are employed for holographic display are imperfect. Any deviation in the locations of the pixels in an SLM, or the error in modulating the phase of the optical waves passing through them, will offset the delicate error compensation mechanism of the error diffusion process. A second factor leading to the poor optical reconstruction is the error that is associated with each of the

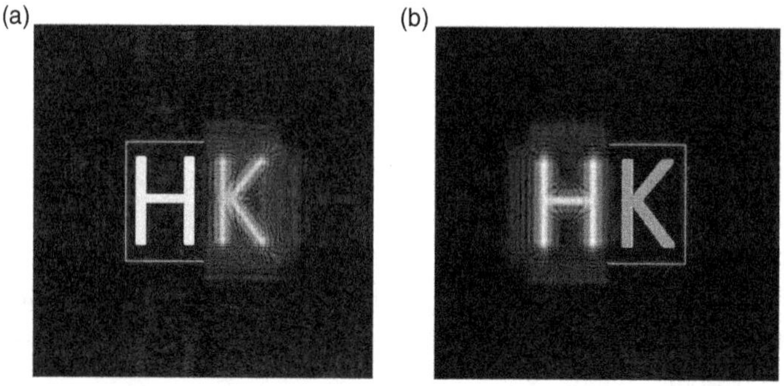

Figure 4.9 Reconstructed image from the BERD hologram: (a) on focused plane at 0.07 m; (b) on focused plane at 0.08 m.

hologram pixels that cannot be fully removed with error diffusion. Grossly speaking, error diffusion is capable of reducing, but not eliminating, the error caused by discarding the magnitude component of a hologram. For an inline hologram, the zero-order beam, together with the optical waves of the error signals, will merge with the reconstructed image of the hologram, resulting in severe degradation of the visual quality.

A simple way of overcoming the abovementioned problem is to add an off-axis reference beam to convert the inline phase-only hologram into an off-axis hologram, so that the reconstructed image can be steered away from the zero-order beam and the error component. Mathematically, the off-axis phase-only hologram is presented as

$$H_{OA}(m,n) = H_P(m,n) \times R(n), \tag{4.21}$$

where $R(n) = e^{i\frac{2\pi}{\lambda}n\delta_d \sin(\theta_R)}$ represents an optical beam inclined at an angle of incidence θ_R along the vertical direction. Since both $H_P(m,n)$ and $R(n)$ are phase-only functions, $H_{OA}(m,n)$ is also a phase-only hologram.

4.6 Localized Error Diffusion

As mentioned in Section 4.5, error diffusion is a recursive process whereby the conversion of a hologram into a phase-only value has to wait for the conversion of all the pixels before it can be completed. The error diffusion process, therefore, cannot be speeded up with parallel processing, and operation at a video rate of 25–30 frames per second is difficult to attain. To overcome this problem, a fast algorithm known as "localized error diffusion" was suggested in [14]. Briefly, a hologram is evenly partitioned into vertical zones each comprising a fixed number of columns. In each zone, modified UERD, which is referred to as localized error diffusion (LERD), is applied to convert the hologram pixels to phase-only values. The LERD enables the error diffusion process in each zone to be conducted in a concurrent manner, so that it can be executed with fast parallel processors such as graphics processing units (GPUs). Details of the method are as follows.

Consider converting a complex-valued hologram $H(m,n)$ into a phase-only hologram. The hologram is uniformly partitioned into K vertical zones, each comprising N columns, as shown in Figure 4.10(a), and each zone is hereafter referred to as a sub-hologram $H_k(m,n)|_{1\leq k\leq K}$. Next, LERD is applied to convert each sub-hologram $H_k(m,n)$ into a phase-only hologram $H_{p;k}(m,n)$.

Similar to the UERD algorithm in Section 4.4, each sub-hologram is processed in a row-by-row manner, progressing from the first row to the last row. In each row, which is referred to as a segment, pixels are scanned from left to right. The magnitude of each visited pixel in a segment is forced to unity, and the phase value is retained. Suppose $P_{m_j;n_j;k}$ denotes the pixel being visited at location (m_j,n_j) in the kth sub-hologram, the error $E_k(m_j,n_j)$ resulting from forcing its magnitude to unity will be

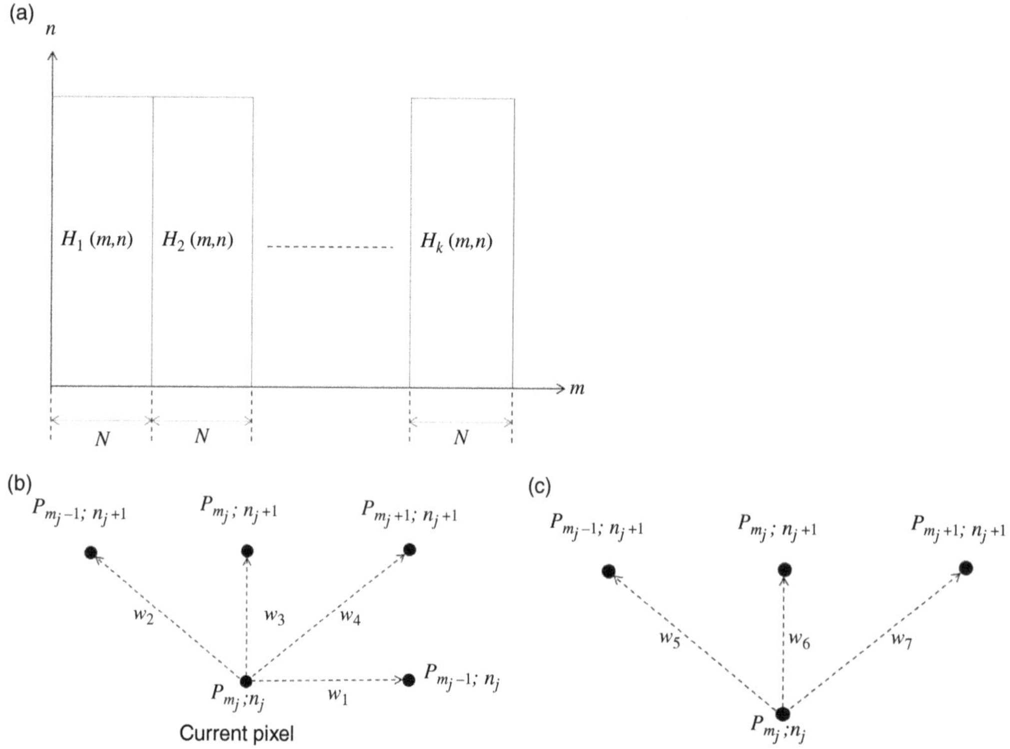

Figure 4.10 (a) Partitioning a hologram into vertical zones. (b) Diffusion of error for a pixel that is not in the last column of a sub-hologram. (c) Diffusion of error for a pixel at the last column of a sub-hologram.

$$E_k(m_j, n_j) = H_k(m_j, n_j) - H_{P;k}(m_j, n_j). \tag{4.22}$$

The error $E_k(m_j, n_j)$ is then added, with preset weighting factors, to the pixels in its neighborhood. For pixels that are not located in the last column of a sub-hologram, the error is added to its four adjacent pixels as shown in Figure 4.10(b):

$$H_k(m_j - 1, n_j) \leftarrow H_k(m_j - 1, n_j) + w_1 E_k(m_j, n_j), \tag{4.23}$$

$$H_k(m_j + 1, n_j + 1) \leftarrow H_k(m_j + 1, n_j + 1) + w_2 E_k(m_j, n_j), \tag{4.24}$$

$$H_k(m_j, n_j + 1) \leftarrow H_k(m_j, n_j + 1) + w_3 E_k(m_j, n_j), \tag{4.25}$$

$$H_k(m_j - 1, n_j + 1) \leftarrow H_k(m_j - 1, n_j + 1) + w_4 E_k(m_j, n_j), \tag{4.26}$$

where w_1 to w_4 are the Floyd–Steinberg error diffusion coefficients.

For pixels in the last column of a sub-hologram, the error is added to the three pixels in the next row, as shown in Figure 4.10(c). Mathematically this is denoted as

$$H_k(m_j + 1, n_j + 1) \leftarrow H_k(m_j + 1, n_j + 1) + w_5 E_k(m_j, n_j), \tag{4.27}$$

$$H_k(m_j, n_j + 1) \leftarrow H_k(m_j, n_j + 1) + w_6 E_k(m_j, n_j), \tag{4.28}$$

$$H_k(m_j - 1, n_j + 1) \leftarrow H_k(m_j - 1, n_j + 1) + w_7(m_j, n_j), \tag{4.29}$$

where $w_5 = \frac{3}{9}$, $w_6 = \frac{5}{9}$, and $w_7 = \frac{1}{9}$.

From Figure 4.10, as well as Eqs. (4.23)–(4.29), in the processing of each row of pixels in a sub-hologram, the error of any pixel in one zone will not propagate to the pixels of other zones on the same row. In another words, the computation performed on any sub-hologram on a given row does not rely on the outcome of other sub-holograms. As such, error diffusion can be conducted concurrently for all the segments in the same row of the sub-holograms, hence enabling parallel processors to be utilized to speed up the conversion process. Suppose the width of each segment is S pixels, and N is the horizontal extent of the hologram, there will be N/S sub-holograms whereby error diffusion can be applied independently to each of them. The computation time can be speeded up by N/S times with the use of parallel processors.

On the downside, the error in the last pixel of a segment will not be distributed to the pixel directly below it; instead it will be distributed to the one on its right. This will lead to heavier distortion at the boundaries between adjacent zones, resulting in repetitive vertical strip patterns in the reconstructed image. This problem can be reduced by applying low-pass filters (for example, using optical filtering with a 4f system) to redistribute the error signal in the phase-only hologram. The low-pass filtering diffuses the error in the last pixel of each segment to its adjacent segment.

4.7 Converting a Complex-Valued Hologram to a Binary Phase-Only Hologram with Direct Binary Search

In Section 3.3.8 it was shown that direct binary search (DBS) [15,16] can be applied to generate a binary phase-only hologram. With minor modification, the DBS can also be used to convert a complex-valued hologram into a phase-only hologram. As the source image is not available, the reconstructed image of the complex-valued hologram, denoted by $H(m, n)$, is computed:

$$I_{ref}(m, n) = H(m, n) * h(m, n; z_0), \tag{4.30}$$

where $h(m, n; z_0)$ is the free space impulse response. The image $I_{ref}(m, n)$ is taken to be the reference image. Similar to the process in Section 3.3.8, the binary phase-only hologram is converted from the original hologram through an iterative process. Let k and $B_k(m, n)$ denote the number of iterations that have been conducted and the corresponding binary phase-only hologram. Initially, at $k = 0$, pixels of $B_{k=0}(m, n)$ are assigned random values of either 1 or 0. A reconstructed image from the binary phase-only hologram can be obtained by

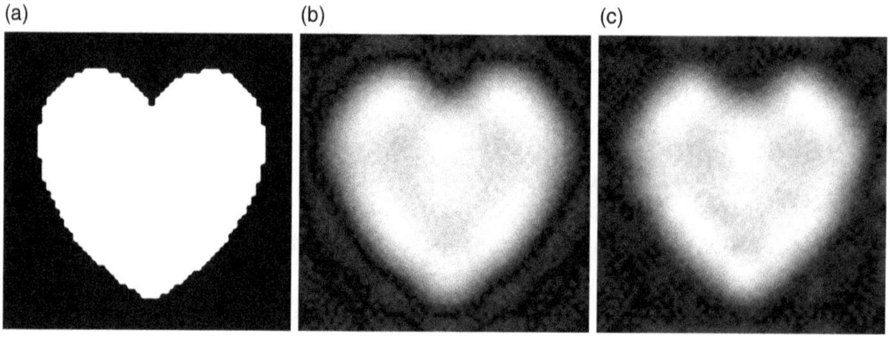

Figure 4.11 (a) Source image "Heart." (b) Reference image (reconstructed image of the complex-valued hologram). (c) Reconstructed image of binary hologram obtain with 10 rounds of DBS.

$$I_{B;k}(m,n) = B_k(m,n) * h^{\dagger}(m,n;z_0), \tag{4.31}$$

where $h^{\dagger}(m,n;z_0)$ is the conjugate of the free space impulse response. In each of the subsequent iterations, a new pixel in the binary hologram is selected along a row-by-row, left-to-right path. When each pixel is visited, its binary value is complemented. A reconstructed image $I_{B;k}(m,n)$ of the updated binary hologram $B_k(m,n)$ on the focused plane is computed using Eq. (4.31). The mean square errors (MSEs) between the reference image and the reconstructed images $I_{B;k}(m,n)$ and $I_{B;k-1}(m,n)$ are determined. If the MSE of the updated hologram is lower than that of the previous hologram at the $(k-1)$th iteration, it will be retained. Otherwise, $B_k(m,n)$ will be replaced by $B_{k-1}(m,n)$. Subsequently, the iteration count k is increased by 1. After all the binary pixels have been evaluated, a binary phase-only hologram will be generated. The quality of the reconstructed image of the binary phase-only hologram generated with a single round of DBS is poor. This problem can be alleviated with multiple rounds of DBS.

A numerical simulation for converting a complex-valued hologram into a binary phase-only hologram with the DBS method is conducted. The source code for conducting a single epoch of DBS is shown in Section 4.8.5. When the program is executed, the user is prompted to enter an integer from 0 to 2. If the input value is 1, the binary hologram is initialized with random values and a new run of the DBS will be conducted. For an input value of 2, the DBS will be conducted based on the binary hologram generated in the previous epoch. As DBS is time consuming, the program can be stopped by the user in the middle of the process. An input value of 1 resumes the DBS after the interruption.

In the simulation, a complex-valued Fresnel hologram $H(m,n)$ of the source image "Heart" in Figure 4.11(a) is adopted. To avoid a long computation time, the sizes of the source image and its hologram are reduced to 64×64 and 128×128, respectively. The axial distance between the hologram and the source image is 0.07 m. The reconstructed image of $H(m,n)$ on the focused plane, as shown in Figure 4.11(b), is computed and taken as the reference image. A binary phase-only hologram is generated after 10 rounds of DBS. The reconstructed image of the binary hologram is shown in Figure 4.11(c). The reconstructed image is similar to the reference image, with a fidelity (in PSNR) of 29.32 dB.

4.8 MATLAB Simulation

This section includes the MATLAB source codes employed to conduct the simulations of the methods reported in this chapter. A summary of the simulation programs (each identified by the name of its main function call) and their correspondence to previous sections are listed in Table 4.1.

Three images are used to demonstrate the methods in this chapter. The first one is a planar image "Apple" that is located at an axial distance 0.08 m from the hologram. This image is used in the CAM simulation. The second image is the double-depth image "HK1" shown in Figure 1.12. The image is evenly divided into a left half-plane and a right half-plane, containing the characters H and K, respectively. The left half-plane is located at a distance of 0.07 m from the hologram plane, while the right half-plane is placed farther away at 0.08 m. This image is adopted in the simulation of the double-phase, UERD, and BERD methods. A smaller image, "Heart," is used in the simulation of the DBS method.

4.8.1 Simulation of Converting a Complex-Valued Hologram into a Phase-Only Hologram with the CAM Method

The source code for this simulation is shown in Table 4.2. A complex-valued digital Fresnel hologram of a planar image is generated. The hologram is then converted into a phase-only hologram with the CAM method. To begin with, the main function "CAM" is called to load the source image file. Next, the function "Holo_gen" is applied to convert the source image into a complex-valued Fresnel hologram.

Table 4.1 Summary of simulation programs in this chapter

Function name	Description of simulation	Associated sections
CAM	Converting a complex-valued hologram into a phase-only hologram with the CAM method	4.2
DP	Converting a complex-valued hologram into a phase-only hologram with the double-phase macroblock method	4.3
UERD	Converting a complex-valued hologram into a phase-only hologram with the UERD method	4.4
BERD	Converting a complex-valued hologram into a phase-only hologram with the BERD method	4.5
DBS	Converting a complex-valued hologram into a binary phase-only hologram with the DBS method	4.6

Table 4.2 Simulation of generating a phase-only hologram with the CAM method

```
function CAM()
  %Read source image and store in a 2-D array
  I=imread('apple.jpg');
  I=im2double(I);

  %Optical setting
  H_DIM=2048;                       %Dimension of hologram
  lamda=520e-9;                     %Wavelength of light=520 nm
  psize=6.2e-6;                     %Pixel size=6.4 µm
  z=0.08;                           %Distance between image and hologram
                                      %plane

  %Parameter beta for modifying the effect of the Bessel function
  beta=input('Enter value of beta ');

  %Generate POH with the CAM method, and return reconstructed image
  H=Holo_gen(I,z,psize,lamda);      %Generate hologram

  %If image is smaller than hologram, compute and display reconstructed image
  Hcam=CAM_GEN(H,beta);             %Convert to hologram CAM phase-only
                                      %hologram
  RES=abs(Holo_recon(Hcam,z,psize,  %Reconstruct CAM POH
    lamda));
  RES=RES/max(RES(:));

  %Show and save reconstructed image
  figure,imshow(mat2gray(RES));
  imwrite(mat2gray(abs(RES)),strcat('RES_CAM_beta',num2str(beta),'.jpg'));

  %Measure similarity between source and reconstructed image in PSNR (db)
  psnr_db=PSNR(I,RES)
end

function HOLO=CAM_GEN(HG,beta)
  %Normalize magnitude of hologram
  MAG=abs(HG);
  MAG=beta*MAG./max(MAG(:));
  ANG=angle(HG);                    %Compute phase component of hologram

  %Apply the Bessel function (m=1) to the magnitude component
  MAG=besselj(1,MAG);

  %==================================================================
  %Store the phase-only hologram to the output array 'HOLO'
  %==================================================================
  HOLO=MAG.*cos(ANG)+1i*MAG.*sin
    (ANG);
end

function H=Holo_gen(I,z,psize,lamda)
  [N M]=size(I(:,:));

  wn=2*pi/lamda;                    %Wave number
```

Table 4.2 (cont.)

```
  %Generate Fresnel zone plate
  z2=z*z;                               %Square of the depth value z
  [kx ky]=meshgrid(-M/2:M/2-1,-N/2:N/2-1);
  kx=kx*psize;                          %Physical horizontal position
  ky=ky*psize;                          %Physical vertical position

  fzp=exp(j*wn*sqrt(kx.^2+ky.^2+z2));
  %Compute hologram of source image
  A=fft2(I);                            %Fourier transform of the hologram
  B=fft2(fzp);                          %Fourier transform of the free space
                                          %impulse response

  H=fftshift(ifft2(A.*B));
end

function IR=Holo_recon(H,z,psize,lamda)
  [N M]=size(H(:,:));                   %Size of hologram

  wn=2*pi/lamda;                        %Wave number

  %Generate Fresnel zone plate
  z2=z*z;                               %Square of the depth value z
  [kx ky]=meshgrid(-M/2:M/2-1,-N/2:N/2-1);
  kx=kx*psize;                          %Physical horizontal position
  ky=ky*psize;                          %Physical vertical position
  r=sqrt(kx.^2+ky.^2+z2);
  fzp_c=exp(-j*wn*r)./r;                %Conjugate of FZP

  A=fft2(H);                            %Fourier transform of the hologram
  B=fft2(fzp_c);                        %Fourier transform of the free space
                                          %impulse response

  %Compute reconstructed image
  IR=fftshift(ifft2(A.*B));
end

function psnr_db=PSNR(IMG,RES)
  [row,col]=size(IMG)
  A=(IMG-RES).^2;
  MSE=sum(A(:))/(row*col);
  psnr_db=10*log10(1/MSE);
end
```

The user is prompted to input a value for the parameter β. Based on Eqs. (4.2) and (4.5), the complex-valued hologram is converted into a phase-only hologram with the function “CAM_GEN.” Subsequently, the reconstructed image at the focused plane is computed from the phase-only hologram with the function “Holo_recon.” The reconstructed image is then displayed and saved. The function “PSNR” is called to determine the fidelity, in PSNR, of the reconstructed image as compared with the source image.

4.8.2 Simulation of Converting a Complex-Valued Hologram into a Phase-Only Hologram with the Double-Phase Macroblock Method

In this simulation, a complex-valued digital Fresnel hologram of a planar image is generated. The source code is shown in Table 4.3.

Table 4.3 Simulation of generating a phase-only hologram with the double-phase macroblock method

```
function DP()
  fname='HK';
  I=imread(strcat(fname,'.jpg'));       %Filename of intensity image
  I=im2double(I);
  [n m]=size(I(:,:));                   %Size of source image,

  N=n*2; M=m*2;                         %size of hologram = Two times size of
                                          %image

  I1=zeros(N,M);
  I1(n-n/2:n+n/2-1,m-m/2:m+m/2-1)=I;%First layer: left half-plane
  I2=I1;                                %Second layer: right half-plane
  I1(:,M/2+1:M)=0;                      %Only left half-plane of source image in
                                          %first layer
  I2(:,1:M/2)=0;                        %Only right half-plane of source image
                                          %in second layer

  psize=6.4*10^-6;                      %Pixel size of hologram=6.4 µm
  lamda=520*10^-9;                      %Wavelength=540 nm

  z1=0.07;                              %Depth of left half-plane of image (first
                                          %layer)
  z2=0.08;                              %Depth of right half-plane of image
                                          %(second layer)

  H1=Holo_gen(I1,z1,psize,lamda);       %Hologram of first layer
  H2=Holo_gen(I2,z2,psize,lamda);       %Hologram of second layer
  H=H1+H2;                              %Hologram=sum of first and second layer
                                          %holograms

  %Convert to phase-only hologram with the double-phase macroblock method
  %mode=0: Column interleave double-phase, mode=1: Checkboard interleave
    %double-phase
  mode=input('Enter 1x2 or 2x2 interleaving method (0 or 1)');
  H_I=Interleave(H,mode);

  %Bandwidth of low-pass filtering
  BW=M/4;
  H=LPF(H_I,BW);

  %Reconstruct double-phase hologram on the two focused planes
  RES1=mat2gray(abs(Holo_recon(H,z1,psize,lamda)));
  RES2=mat2gray(abs(Holo_recon(H,z2,psize,lamda)));

  %Show and save reconstructed images
  figure,imshow(RES1);
  imwrite(RES1,strcat('RES_DP_IN',num2str(mode),'_D1.jpg'));
```

Table 4.3 (cont.)

```
  figure,imshow(RES2);
  imwrite(RES2,strcat('RES_DP_IN',num2str(mode),'_D2.jpg'));

end

function H=Holo_gen(I,z,psize,lamda)

[N M]=size(I(:,:));
wn=2*pi/lamda;                          %Wave number

%Generate Fresnel zone plate
z2=z*z;                                 %Square of the depth value z
[kx ky]=meshgrid(-M/2:M/2-1,-N/2:N/2-1);

kx=kx*psize;                            %Physical horizontal position
ky=ky*psize;                            %Physical vertical position
fzp=exp(j*wn*sqrt(kx.^2+ky.^2+z2));

%Compute hologram of source image
A=fft2(I);                              %Fourier transform of the hologram
B=fft2(fzp);                            %Fourier transform of the free space
                                          %impulse response

H=fftshift(ifft2(A.*B));
end

function HOLO=Interleave(H,mode)
  [N,M]=size(H(:,:));

  %Normalize magnitude of hologram
  HOLO=zeros(N,M);
  MAG=abs(H);
  MAG=MAG./max(MAG(:));

  t1=angle(H);                          %Phase component of hologram
  t2=acos(MAG);                         %Phase component corresponding to
                                          %magnitude of hologram
  r1=t1+t2;                             %First phase angle of double-phase
                                          %representation

  r2=t1-t2;                             %Second phase angle of double-phase
                                          %representation

  if mode==0
    for i=1:N
      for j=1:2:M
        HOLO(i,j)= cos(r1(i,j))+1i*sin(r1(i,j));
        HOLO(i,j+1)=cos(r2(i,j+1))+1i*sin(r2(i,j+1));
      end
    end
  else
    for i=1:N
      for j=1:2:M
        if mod(i,2)==0
          HOLO(i,j)= cos(r1(i,j))+1i*sin(r1(i,j));
          HOLO(i,j+1)=cos(r2(i,j+1))+1i*sin(r2(i,j+1));
```

Table 4.3 (cont.)

```
          else
             HOLO(i,j+1)= cos(r1(i,j+1))+1i*sin(r1(i,j+1));
             HOLO(i,j)=cos(r2(i,j))+1i*sin(r2(i,j));
          end
        end
      end
   end
end

function IR=Holo_recon(H,z,psize,lamda)
   [N M]=size(H(:,:));                     %Size of hologram
   wn=2*pi/lamda;                          %Wave number
   %Generate Fresnel zone plate
   z2=z*z;                                 %Square of the depth value z
   [kx ky]=meshgrid(-M/2:M/2-1,-N/2:N/2-1);
   kx=kx*psize;                            %Physical horizontal position
   ky=ky*psize;                            %Physical vertical position
   r=sqrt(kx.^2+ky.^2+z2);
   fzp_c=exp(-j*wn*r)./r;                  %Conjugate of FZP

   A=fft2(H);                              %Fourier transform of the hologram
   B=fft2(fzp_c);                          %Fourier transform of the free space
                                             %impulse response

   IR=fftshift(ifft2(A.*B));               %Compute reconstructed image
end

function HL=LPF(H,BW)
   [N,M]=size(H(:,:));
   if BW>0                                 %No filtering for BW=negative
      F_H=fft2(H);                         %Convert hologram to frequency domain

      %Low-pass filtering
      F_H(:,BW:N-BW)=0;
      F_H(BW:M-BW,:)=0;
      HL=ifft2(F_H);                       %Convert filtered hologram to spatial
                                             %domain

   end
end
```

The hologram is then converted into a phase-only hologram with the double-phase macroblock method. To begin with, the main function "DP," is called to load the source image file. Next, the function "Holo_gen" is applied to convert the source image into a complex-valued Fresnel hologram. The user will be prompted to select either the 1×2 or the 2×2 macro-pixel topology, and the complex-valued hologram will be converted into a phase-only hologram with the function "Interleave." Subsequently, the reconstructed image at the focused plane will be computed from the phase-only hologram with the function "Holo_recon." The reconstructed image is then displayed and saved in a file.

4.8.3 Simulation of Converting a Complex-Valued Hologram into a Phase-Only Hologram with the UERD Method

The source code of this simulation is shown in Table 4.4. The main function "UERD" is executed to load the double-depth image "HK1," from which a complex-valued Fresnel hologram is generated with the function "Holo_gen." Next, the function "UERD_convert" is applied to convert the complex-valued hologram into a phase-only hologram. In this process, hologram pixels are scanned from the first row to the last row, and in each row in a left-to-right direction. The magnitude of each visited pixel is set to unity, and the error is distributed to its four neighbors. Upon completion of the scanning process, a phase-only hologram is obtained. Subsequently, the reconstructed images of the phase-only hologram on two focused planes are computed with the function "Holo_recon." The reconstructed images is then displayed and saved.

Table 4.4 Converting a complex-valued hologram into a phase-only hologram with the UERD method

```
function UERD()
  fname='HK1';
  I=imread(strcat(fname,'.jpg'));      %Filename of intensity image
  I=im2double(I);
  [n m]=size(I(:,:));                  %Size of source image,

  N=n*2; M=m*2;                        %size of hologram = Two times size of image

  I1=zeros(N,M);
  I1(n-n/2:n+n/2-1,m-m/2:m+m/2-1)=I;   %First layer: left half-plane
  I2=I1;                               %Second layer: right half-plane
  I1(:,M/2+1:M)=0;                     %Only left half-plane of source image in
                                         %first layer
  I2(:,1:M/2)=0;                       %Only right half-plane of source image
                                         %in second layer

  psize=6.4*10^-6;                     %Pixel size of hologram=6.4 µm
  lamda=520*10^-9;                     %Wavelength=540 nm

  z1=0.07;                             %Depth of left half-plane of image (first
                                         %layer)
  z2=0.08;                             %Depth of right half-plane of image
                                         %(second layer)

  H1=Holo_gen(I1,z1,psize,lamda);      %Hologram of first layer
  H2=Holo_gen(I2,z2,psize,lamda);      %Hologram of second layer
  H=H1+H2;                             %Hologram=sum of first and second layer
                                         %holograms

  %Reconstruct holograms
  H_UERD=UERD_convert(H);
  RES1=abs(Holo_recon(H_UERD,z1,psize,lamda));
  RES2=abs(Holo_recon(H_UERD,z2,psize,lamda));

  %Show and save reconstructed images
  figure,imshow(mat2gray(RES1));
```

Table 4.4 (cont.)

```
  imwrite(mat2gray(abs(RES1)),'RES_UERD_D1.jpg');
  figure,imshow(mat2gray(RES2));
  imwrite(mat2gray(abs(RES2)),'RES_UERD_D2.jpg');

end

function H=Holo_gen(I,z,psize,lamda)
  [N M]=size(I(:,:));

  wn=2*pi/lamda;                          %Wave number

  %Generate Fresnel zone plate
  z2=z*z;                                 %Square of the depth value z
  [kx ky]=meshgrid(-M/2:M/2-1,-N/2:N/2-1);
  kx=kx*psize;                            %Physical horizontal position
  ky=ky*psize;                            %Physical vertical position
  fzp=exp(j*wn*sqrt(kx.^2+ky.^2+z2));

  %Compute hologram of source image
  A=fft2(I);                              %Fourier transform of the hologram
  B=fft2(fzp);                            %Fourier transform of the free space
                                            %impulse response

 H=fftshift(ifft2(A.*B));
end

function HOLO=UERD_convert(H)
[N,M]=size(H(:,:));
  H_DIM=M;                                %Assume a square hologram with N=M=H_DIM
  C=H./max(abs(H(:)));

  %Extract real and imaginary parts of the complex-valued hologram
  x=real(C);
  y=imag(C);

  %=================================================================
  %Convert complex-valued hologram to POH with error diffusion
  %=================================================================
  %Define Floyd-Steinberg error diffusion coefficients
  w1=7/16;
  w2=3/16;
  w3=5/16;
  w4=1/16;

  for i = 1:H_DIM-1                       %Visit hologram pixels row by row
    for j = 2:H_DIM-1                     %From left to right in each row
      x0=x(i,j);                          %Get real part of current pixel
      y0=y(i,j);                          %Get imaginary part of current pixel
      ang=angle(x0+1i*y0);                %Compute angle of current pixel

      %Force magnitude of current pixel to unity
      x(i,j)=cos(ang);
      y(i,j)=sin(ang);
```

Table 4.4 (cont.)

```
      %Compute error of converting complex to phase-only value
      err_x = x0-x(i,j);                %Error of real part
      err_y = y0-y(i,j);                %Error of imaginary part

      %Diffuse error of real part to the four neighboring pixels
      x(i,j+1)=x(i,j+1)+ w1*err_x;
      x(i+1,j-1)=x(i+1,j-1)+ w2*err_x;
      x(i+1,j)=x(i+1,j)+ w3*err_x;
      x(i+1,j+1)=x(i+1,j+1)+ w4*err_x;

      %Diffuse error of imaginary part to the four neighboring pixels
      y(i,j+1)=y(i,j+1)+ w1*err_y;
      y(i+1,j-1)=y(i+1,j-1)+ w2*err_y;
      y(i+1,j)=y(i+1,j)+ w3*err_y;
      y(i+1,j+1)=y(i+1,j+1)+ w4*err_y;
    end
  end

  for i = 1:H_DIM
    x(i,1)=x(i,2);
    x(i,H_DIM)=x(i,H_DIM-1);
  end
  for i = 1:H_DIM
    x(H_DIM,i)=x(H_DIM-1,i);
    x(1,i)=x(2,i);
 end

  %=================================================================
  %Store the phase-only hologram to the output array 'HOLO'
  %=================================================================
  r=angle(x+1i*y);
  HOLO=cos(r)+1i*sin(r);
end

function IR=Holo_recon(H,z,psize,lamda)
  [N M]=size(H(:,:));                   %Size of hologram

  wn=2*pi/lamda;                        %Wave number
  %Generate Fresnel zone plate
  z2=z*z;                               %Square of the depth value z
  [kx ky]=meshgrid(-M/2:M/2-1,-N/2:N/2-1);
  kx=kx*psize;                          %Physical horizontal position
  ky=ky*psize;                          %Physical vertical position
  r=sqrt(kx.^2+ky.^2+z2);
  fzp_c=exp(-j*wn*r)./r;                %Conjugate of FZP

  A=fft2(H);                            %Fourier transform of the hologram
  B=fft2(fzp_c);                        %Fourier transform of the free space
                                          %impulse response
  IR=fftshift(ifft2(A.*B));             %Compute reconstructed image
end
```

4.8.4 Simulation of Converting a Complex-Valued Hologram into a Phase-Only Hologram with the BERD Method

In this simulation, a digital phase-only Fresnel hologram of a double-depth image is generated. The source code is shown in Table 4.5. To begin, in the main function "BERD" the double-depth image "HK1" is loaded, from which a complex-valued hologram is generated with the function "Holo_gen." Next, the function "BERD_convert" is applied to convert the complex-valued hologram into a phase-only hologram. The process is similar to the UERD method described in Section 4.8.3, with the exception that the direction of scanning for the even and the odd rows is reversed. Subsequently, the reconstructed image at the focused plane will be computed from the phase-only hologram with the function "Holo_recon." The reconstructed image is then displayed and saved.

Table 4.5 Converting a complex-valued hologram into a phase-only hologram with the BERD method

```
function BERD()
   fname='HK1';
   I=imread(strcat(fname,'.jpg'));          %Filename of intensity image
   I=im2double(I);
   [n m]=size(I(:,:));                      %Size of source image,
   N=n*2; M=m*2;                            %size of hologram = Two times size of
                                              %image

   I1=zeros(N,M);
   I1(n-n/2:n+n/2-1,m-m/2:m+m/2-1)=I;       %First layer: left half-plane
   I2=I1;                                   %Second layer: right half-plane
   I1(:,M/2+1:M)=0;                         %Only left half-plane of source image in
                                              %first layer
   I2(:,1:M/2)=0;                           %Only right half-plane of source image
                                              %in second layer

   psize=6.4*10^-6;                         %Pixel size of hologram=6.4 µm
   lamda=520*10^-9;                         %Wavelength=540 nm

   z1=0.07;                                 %Depth of left half-plane of image (first
                                              %layer)
   z2=0.08;                                 %Depth of right half-plane of image
                                              %(second layer)

   H1=Holo_gen(I1,z1,psize,lamda);          %Hologram of first layer
   H2=Holo_gen(I2,z2,psize,lamda);          %Hologram of second layer
   H=H1+H2;                                 %Sum of first and second layer holograms

   %Reconstruct holograms
   H_UERD=BERD_convert(H);
   RES1=abs(Holo_recon(H_UERD,z1,psize,lamda));
   RES2=abs(Holo_recon(H_UERD,z2,psize,lamda));

   %Show and save reconstructed images
   figure,imshow(mat2gray(RES1));
   imwrite(mat2gray(abs(RES1)),'RES_BERD_D1.jpg');
   figure,imshow(mat2gray(RES2));
```

Table 4.5 (cont.)

```
  imwrite(mat2gray(abs(RES2)),'RES_BERD_D2.jpg');
end

function H=Holo_gen(I,z,psize,lamda)
  [N M]=size(I(:,:));

  wn=2*pi/lamda;                          %Wave number

  %Generate Fresnel zone plate
  z2=z*z;                                 %Square of the depth value z
  [kx ky]=meshgrid(-M/2:M/2-1,-N/2:N/2-1);
  kx=kx*psize;                            %Physical horizontal position
  ky=ky*psize;                            %Physical vertical position
  fzp=exp(j*wn*sqrt(kx.^2+ky.^2+z2));

  %Compute hologram of source image
  A=fft2(I);                              %Fourier transform of the hologram
  B=fft2(fzp);                            %Fourier transform of the free space
                                            %impulse response
  H=fftshift(ifft2(A.*B));
end
function HOLO=BERD_convert(H)

  [N M]=size(H(:,:));                     %Size of hologram
  H_DIM=M;                                %Assume a square hologram with
                                            %M=N=H_DIM
  C=H./max(abs(H(:)));
  %Extract real and imaginary parts of the complex-valued hologram
  x=real(C);
  y=imag(C);

  %=================================================================
  %Convert complex-valued hologram to POH with error diffusion
  %=================================================================
  %Define Floyd-Steinberg error diffusion coefficients
  w1=7/16;
  w2=3/16;
  w3=5/16;
  w4=1/16;

  for i = 1:H_DIM-1                       %Visit hologram pixels row by row
    %Error diffusion of odd (1,3,5, …) rows
    if mod(i,2) == 1
      for j = 2:H_DIM-1                   %From left to right in odd row
          x0=x(i,j);                      %Get real part of current pixel
          y0=y(i,j);                      %Get imaginary part of current pixel
          ang=angle(x0+1i*y0);            %Compute angle of current pixel

          %Force magnitude of current pixel to unity
          x(i,j)=cos(ang);
          y(i,j)=sin(ang);

          %Compute error of converting complex to phase-only value
```

Table 4.5 (cont.)

```
            err_x = x0-x(i,j);              %Error of real part
            err_y = y0-y(i,j);              %Error of imaginary part
            %Diffuse error of real part to the four neighboring pixels
            x(i,j+1)=x(i,j+1) + w1*err_x;
            x(i+1,j-1)=x(i+1,j-1) + w2*err_x;
              x(i+1,j)=x(i+1,j) + w3*err_x;
            x(i+1,j+1)=x(i+1,j+1) + w4*err_x;

            %Diffuse error of imaginary part to the four neighboring pixels
            y(i,j+1)=y(i,j+1) + w1*err_y;
            y(i+1,j-1)=y(i+1,j-1) + w2*err_y;
            y(i+1,j)=y(i+1,j) + w3*err_y;
            y(i+1,j+1)=y(i+1,j+1) + w4*err_y;
          end
      else
          %Error diffusion of even (2,4,6, …) rows
          for j = H_DIM-1:-1:2                %From right to left in even row
            x0=x(i,j);                        %Get real part of current pixel
            y0=y(i,j);                        %Get imaginary part of current pixel
            ang=angle(x0+1i*y0);              %Compute angle of current pixel

            %Force magnitude of current pixel to unity
            x(i,j)=cos(ang);
            y(i,j)=sin(ang);

            %Compute error of converting complex to phase-only value
            err_x = x0-x(i,j);                %Error of real part
            err_y = y0-y(i,j);                %Error of imaginary part

            %Diffuse error of real part to the four neighboring pixels
            x(i,j-1)=x(i,j-1) + w1*err_x;
            x(i+1,j+1)=x(i+1,j+1) + w2*err_x;
            x(i+1,j)=x(i+1,j) + w3*err_x;
            x(i+1,j-1)=x(i+1,j-1) + w4*err_x;

            %Diffuse error of imaginary part to the four neighboring pixels
            y(i,j-1)=y(i,j-1) + w1*err_y;
            y(i+1,j+1)=y(i+1,j+1) + w2*err_y;
            y(i+1,j)=y(i+1,j) + w3*err_y;
            y(i+1,j-1)=y(i+1,j-1) + w4*err_y;
          end
      end
  end

  for i = 1:H_DIM
    x(i,1)=x(i,2);
    x(i,H_DIM)=x(i,H_DIM-1);
  end
  for i = 1:H_DIM
    x(H_DIM,i)=x(H_DIM-1,i);
    x(1,i)=x(2,i);
  end
```

Table 4.5 (cont.)

```
  %=================================================================
  %Store the phase-only hologram to the output array 'HOLO'
  %=================================================================
  r=angle(x+1i*y);
  HOLO=cos(r)+1i*sin(r);
end

function IR=Holo_recon(H,z,psize,lamda)
  [N M]=size(H(:,:));                  %Size of hologram
  wn=2*pi/lamda;                       %Wave number

  %Generate Fresnel zone plate
  z2=z*z;                              %Square of the depth value z
  [kx ky]=meshgrid(-M/2:M/2-1,-N/2:N/2-1);
  kx=kx*psize;                         %Physical horizontal position
  ky=ky*psize;                         %Physical vertical position
  r=sqrt(kx.^2+ky.^2+z2);
  fzp_c=exp(-j*wn*r)./r;               %Conjugate of FZP

  A=fft2(H);                           %Fourier transform of the hologram
  B=fft2(fzp_c);                       %Fourier transform of the free space
                                         %impulse response
  IR=fftshift(ifft2(A.*B));            %Compute reconstructed image
end
```

4.8.5 Simulation of Converting a Complex-Valued Hologram into a Binary Phase-Only Hologram with the DBS Method

In this simulation, a digital phase-only Fresnel hologram of a planar image is generated. The source code is shown in Table 4.6. In the main function “DBS,” the image “Heart” is loaded, from which a complex-valued hologram is generated with the function “Holo_gen.” Next, the user is prompted to enter a value of 0, 1, or 2. The input value indicates whether the DBS should be conducted as a new run (0), resume from the previous run (1), or as a new epoch based on the binary hologram obtained in the previous run (2). Upon completion of a complete round of DBS, or when the process is stopped by the user, the reconstructed image at the focused plane is computed from the phase-only hologram with the function “Holo_recon.” The reconstructed image is then displayed and saved.

4.9 Summary

In Chapter 3 a number of methods were described for generating a phase-only hologram of an object. However, these methods are not applicable if the source image of the object is not present and only its hologram is available. Such a situation happens if a hologram is directly captured from a physical object

Table 4.6 Converting a complex-valued hologram into a binary phase-only hologram with the DBS method

```
function DBS()
close all
  %Read source image and store in a 2-D array
  I0=imread('Heart.jpg');
  I0=im2double(I0);
  [n m]=size(I0(:,:));                 %size of source image

  [n %Optical setting
  [n H_DIM=m*2;                        %Dimension of hologram
  [n lamda=520e-9;                     %Wavelength of light=520 nm
  [n psize=6.4e-6;                     %Pixel size=6.4 µm
  z=0.07;                              %Distance between image and hologram
                                         %plane

  %Copy source image into a larger canvas that is identical
  %in size as the hologram
  I=zeros(H_DIM,H_DIM);
  B=randi([0 1],H_DIM,H_DIM);          %Initial binary hologram with random
                                         %values
  sx=(H_DIM-m)/2;ex=(H_DIM+m)/2-1;
  sy=(H_DIM-n)/2;ey=(H_DIM+n)/2-1;
  I(sy:ey,sx:ex)=I0;

  %Generate POH with the CAM method, and return reconstructed image
  H=Holo_gen(I,z,psize,lamda);         %Generate hologram

  %If image is smaller than hologram, compute and display reconstructed image
  RES0=abs(Holo_recon(H,z,psize,       %Reconstruct image of original
    lamda));                             %hologram
  RES=mat2gray(RES0(sy:ey,sx:ex));

  mode=input('New run or Resume or Repeat (0/1/2)? ');
  if mode==0                           %Run DBS for the first time, i.e., a new
                                         %run
   Bp=B*pi;
   RES1_0=abs(Holo_recon(Bp,z,psize,lamda)); %Reconstruct image of binary
     hologram
   RES1=mat2gray(RES1_0(sy:ey,sx:ex));
   FD_previous=PSNR(RES,RES1);
   p_row=1;
   p_col=1;
  end
  if mode==1 || mode==2                %Resume previous DBS process
   load('binary.mat');
   load('P.mat');
   if mode==2                          %Start another epoch of DBS, based on
                                         %previous results
     p_row=P(1,1);                     %Resume previous epoch of DBS from the
                                         %last stop point
     p_col=P(1,2);
```

Table 4.6 (cont.)

```
    else
      p_row=1;
      p_col=1;
  end
    FD_previous=P(1,3);
    Bp=B*pi;
   RES2_0=abs(Holo_recon(Bp,z,psize,%Reconstruct image of original
     lamda));                        %hologram
    RES2=mat2gray(RES2_0(sy:ey,sx:ex));
    figure
    imshow(RES2);
   end
   f = waitbar(p_row/H_DIM,'Please wait ...');
   for row=p_row:H_DIM
     for col=p_col:H_DIM
       B1=B;
       B1(row,col)=1-B1(row,col);
       Bp=B1*pi;
       RES2_0=abs(Holo_recon(Bp,z,psize,lamda)); %Reconstruct image of
 %original hologram
       RES2=mat2gray(RES2_0(sy:ey,sx:ex));
       FD_current=PSNR(RES,RES2);       %PSNR of current binary hologram

      %Replace the hologram with the current one if the PSNR of its reconstructed
 %image is increased
       if FD_current>FD_previous
         B=B1;
         FD_previous=FD_current
       end
     end
     waitbar(row/H_DIM,f);

     %Save result of current iteration
     save('binary.mat','B');            %Save the binary hologram
     P=zeros(1,3);
     P(1,1)=p_row;
     P(1,2)=p_col;
     P(1,3)=FD_previous;
     save('P.mat','P');                 %Save current parameters
   end
   close(f);

   Bp=B1*pi;
   RES2_0=abs(Holo_recon(Bp,z,psize,%Reconstruct image of original
 lamda));                            %hologram
   RES2=mat2gray(RES2_0(sy:ey,sx:ex));
   figure,imshow(RES2);                %Show reconstructed image
   imwrite(RES,'RES.jpg');             %Save reconstructed image
end
```

Table 4.6 (cont.)

```
function H=Holo_gen(I,z,psize,lamda)
   [N M]=size(I(:,:));

   wn=2*pi/lamda;                       %Wave number
   %Generate Fresnel zone plate
   z2=z*z;                              %Square of the depth value z
   [kx ky]=meshgrid(-M/2:M/2-1,-N/2:N/2-1);
  kx=kx*psize;                          %Physical horizontal position
   ky=ky*psize;                         %Physical vertical position
   fzp=exp(j*wn*sqrt(kx.^2+ky.^2+z2));

   %Compute hologram of source image
   A=fft2(I);                           %Fourier transform of the hologram
   B=fft2(fzp);                         %Fourier transform of the free space
                                          %impulse response
   H=fftshift(ifft2(A.*B));
end

function IR=Holo_recon(H,z,psize,lamda)
   [N M]=size(H(:,:));                  %Size of hologram

   wn=2*pi/lamda;                       %Wave number
   %Generate Fresnel zone plate
   z2=z*z;                              %Square of the depth value z
   [kx ky]=meshgrid(-M/2:M/2-1,-N/2:N/2-1);
   kx=kx*psize;                         %Physical horizontal position
   ky=ky*psize;                         %Physical vertical position
   r=sqrt(kx.^2+ky.^2+z2);
   fzp_c=exp(-j*wn*r)./r;               %Conjugate of FZP

   A=fft2(H);                           %Fourier transform of the hologram
   B=fft2(fzp_c);                       %Fourier transform of the free space
                                          %impulse response

   %Compute reconstructed image
   IR=fftshift(ifft2(A.*B));
end

function psnr_db=PSNR(IMG,RES)
   [row,col]=size(IMG);
   A=(IMG-RES).^2;
   MSE=sum(A(:))/(row*col);
   psnr_db=10*log10(1/MSE);
end
```

(for example, applying phase-shifting holography) instead of generated from a numerical graphic model. This chapter describes six methods for converting a complex-valued hologram into a phase-only hologram. The first two methods, CAM and double-phase methods, convert a complex-valued hologram into a pure phase representation. When the latter is displayed on an SLM with suitable optical filtering, a visual 3-D image is reconstructed. The third to fifth methods apply different variants of the Floyd–Steinberg error diffusion algorithm to convert a complex-valued hologram into a continuous tone phase-only hologram. Among these three error diffusion methods, BERD results in the best reconstructed image, while the LERD method can be implemented with parallel computing devices such as GPUs. The last method, DBS, converts a complex-valued hologram into a binary phase-only hologram through an iterative process. Compared with other methods, the quality of the reconstructed image is generally poor unless more iterations are performed at the expense of longer computation time. A phase-only hologram generated by error diffusion or DBS can be displayed directly with a phase-only SLM without additional optical processing.

Exercises

E.4.1 A hologram is converted into a phase-only hologram and reconstructed with the CAM method. Based on the MATLAB code in Section 4.8.1, write a program to compute the difference (in PSNR) between the reconstructed image of the original hologram and the one obtained with the CAM method, as a function of β.

E.4.2 Further to E.4.1, plot the PSNR versus β, and determine the optimal value of β that results in the maximum PSNR value?

E.4.3 Referring to Eq. (4.6), determine the double-phase representation of the complex number $(0.5 + i0.3)$.

E.4.4 In Eq. (4.7), a binary double-phase hologram can be implemented by quantizing the phase components $\theta_1(x, y)$ and $\theta_2(x, y)$ into a bi-level representation. Based on the MATLAB code in Section 4.8.2, generate a binary double-phase hologram with the 2×2 macro-pixel topology. Evaluate the visual quality and the PSNR of the reconstructed image of the binary double-phase hologram compared with the one obtained with the original complex-valued hologram.

E.4.5 Repeat E.4.4 with a binary double-phase hologram that generated with the 1×2 macro-pixel topology.

E.4.6 A pixel in a complex-valued hologram has a value of $(1.2 + i2.0)$. Determine the error resulting from setting the magnitude of the pixel value to unity. The error is distributed to the four neighboring pixels with the Floyd–Steinberg error diffusion algorithm. What will be the error imparted to each pixel?

References

[1] P.W.M. Tsang, T.-C. Poon, and J.-P. Liu, "Fast extended depth-of-field reconstruction for complex holograms using block partitioned entropy minimization," *Appl. Sci.* vol. 8, no. 5, p. 830 (2018).

[2] Z. Ren, N. Chen, and E. Lam, "Extended focused imaging and depth map reconstruction in optical scanning holography," *Appl. Opt.*, vol. 55, pp. 1040–1047 (2016).

[3] F. Zhao, X. Qu, X. Zhang, T. Poon, T. Kim, Y. Kim, and J. Liang, "Solving inverse problems for optical scanning holography using an adaptively iterative shrinkage-thresholding algorithm," *Opt. Express*, vol. 20, pp. 5942–5954 (2012).

[4] P.W.M. Tsang, K. Cheung, T. Kim, Y. Kim, and T.-C. Poon, "Fast reconstruction of sectional images in digital holography," *Opt. Lett.*, vol. 36, pp. 2650–2652 (2011).

[5] J. Davis, D. Cottrell, J. Campos, M. Yzuel, and I. Moreno, "Encoding amplitude information onto phase-only filters," *Appl. Opt.*, vol. 38, pp. 5004–5013 (1999).

[6] J. Liu and X. Li, "Complex amplitude modulation in real time holographic computation," in *Imaging and Applied Optics 2014, OSA Technical Digest* (Optical Society of America, 2014), paper SM4F.1.

[7] X. Li, J. Liu, J. Jia, Y. Pan, and Y. Wang, "3D dynamic holographic display by modulating complex amplitude experimentally," *Opt. Express*, vol. 21, pp. 20577–20587 (2013).

[8] C. Hsueh and A. Sawchuk, "Computer-generated double-phase holograms," *Appl. Opt.*, vol. 17, pp. 3874–3883 (1978).

[9] J.M. Florence and R.D. Juday, "Full-complex spatial filtering with a phase mostly DMD," in *Proc. SPIE 1558, Wave Propagation and Scattering in Varied Media II*, San Diego, 1991.

[10] V. Arrizón and D. Sánchez-de-la-Llave, "Double-phase holograms implemented with phase-only spatial light modulators: performance evaluation and improvement," *Appl. Opt.*, vol. 41, pp. 3436–3447 (2002).

[11] P.W.M. Tsang and T.-C. Poon, "Novel method for converting digital Fresnel hologram to phase-only hologram based on bidirectional error diffusion," *Opt. Express*, vol. 21, pp. 23680–23686 (2013).

[12] R.W. Floyd and L. Steinberg, "An adaptive algorithm for spatial grey scale," *Proc. Soc. Info. Disp.*, vol. 17, pp. 75–77 (1976).

[13] G. Yang, S. Jiao, J. Liu, T. Lei, and X. Yuan, "Error diffusion method with optimized weighting coefficients for binary hologram generation," *Appl. Opt.*, vol. 58, pp. 5547–5555 (2019).

[14] P.W.M. Tsang, A. Jiao, and T.-C. Poon, "Fast conversion of digital Fresnel hologram to phase-only hologram based on localized error diffusion and redistribution," *Opt. Express*, vol. 22, pp. 5060–5066 (2014).

[15] J.-P. Liu, C. Yu, and P.W.M. Tsang, "Enhanced direct binary search algorithm for binary computer-generated Fresnel holograms," *Appl. Opt.*, vol. 58, pp. 3735–3741 (2019).

[16] T. Leportier and M. C. Park, "Generation of binary holograms for deep scenes captured with a camera and a depth sensor," *Opt. Eng.*, vol. 56, art. no. 013107 (2017).

5 Applications of Phase-Only Hologram in Display, Holographic Encryption, and Steganography

5.1 Introduction

Digital holography has indeed led to numerous advancement of the classical, analog holographic technology that only permits a hologram to be permanently recorded onto a photographic film. In digital holography, a hologram can be captured from a real object. It can also be numerically generated as an array of numbers that can be stored as digital data, processed through computation, and distributed via digital communication links. In general, the primary purpose of holograms is to display 3-D images. Hence, a digital hologram in digital data form will not be of much practical use if it cannot be visually seen as a 3-D image. This is one of the major disadvantages of a digital hologram compared with the optical hologram, which can be readily captured with our eyes. However, a digital hologram can have different applications apart from generating 3-D images. In fact, recent research has shown that a digital hologram can be utilized in protecting sensitive data (a task referred to as cryptography), or in steganography for embedding large amount of additional data. This chapter describes some of the important applications of digital phase-only holograms in 3-D display, holographic cryptography, and steganography.

5.2 Holographic Projection and Display

Modern computing technologies have enabled digital holograms to be generated with a commodity computer at video rate of over 25 frames per second. One may wonder why a holographic display unit is unavailable on the consumer market. The bottleneck in the popularity of holographic displays is caused by two major problems. First, a hologram is a complex-valued image composed of magnitude and phase components, and existing display devices can only present either the magnitude or the phase component. Second, holographic images comprise high-frequency fringes with the spatial frequency of optical waves. For the visible light spectrum, the wavelength of light is within the range [380 nm, 720 nm]. As such, the spatial period of fringes in a hologram of acceptable quality has to be under 10 μm, which is significantly less than the pixel size (roughly in the range 50–200 μm) of consumer-grade electronic displays (such as computer monitors and mobile phones).

The guiding objective of this book is to describe different methods developed to date for producing phase-only holograms, addressing the first problem mentioned above. For the second problem, so far digital holograms can only be displayed on special high-resolution devices that are commonly referred to as spatial light modulators (SLMs). Current SLMs are relatively small, in the order of a few centimeters, but are expensive. It will take years for them to evolve to a comparable size to televisions or computer monitors and to become a consumer commodity. Nevertheless, an SLM is still an effective tool for advancing the frontier of holographic technology. This section will describe the experimental setup and the important factors in displaying holograms on an SLM.

5.2.1 Spatial Light Modulator

"Spatial light modulator" is a general term for devices that are capable of modulating the amplitude, phase, or polarization of light. A typical SLM consists of a 2--D rectangular array of cells (pixels). The optical properties of light passing through or reflecting from each cell can be individually controlled by applying an external electrical signal. There are two major types of SLMs. The first one is based on translucent liquid crystal (LC) devices that modulates the light passing through them. The second type, SLM with liquid crystal on silicon (LCoS), reflect and modulate the light that impinges on it. The quality of an SLM is evaluated with a set of parameters obtained through a number of measurements. These parameters, listed and described in Table 5.1, are crucial in a holographic display system.

For effective holographic display, an SLM should have a large dimension, small pixel pitch, short response time, high light utilization efficiency, and large fill factor.

Table 5.1 Major specification of an SLM

Parameter	Description
Dimension	Number of rows and columns of pixels in the SLM
Pixel pitch	The distance between adjacent pixels. The smaller the distance, the higher the resolution of the SLM
Response time	The time taken for a pixel to switch from 10% to 90% in a range of 2π radians
Light utilization efficiency	Ratio of the average reflected/transmitted light level to the incidence light level
Fill factor	Surface area of the SLM that can be actively utilized. The interpixel gap (the unused area separating a pair of adjacent pixels) decreases the fill factor, thus lowering the diffraction efficiency.

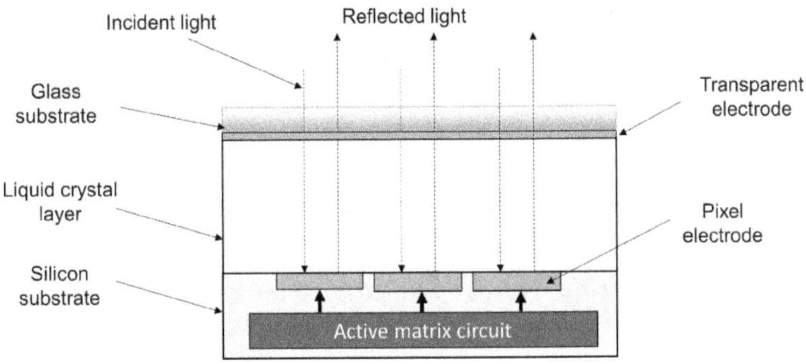

Figure 5.1 Structure of an LCoS SLM.

Liquid crystal on silicon devices are commonly used in modern overhead projectors, and it is a preferable technology compared to LC devices due to its larger fill factor and higher light utilization efficiency. There are numerous LCoS-based SLM products available on the market, with large deviations in their technical specifications. Some of the popular suppliers include, but are not limited to, Hamamatsu Photonics, Holoeye Photonics AG, and the Jasper Display Corp.

A typical LCoS SLM device consists of a layer of liquid crystal material sandwiched between silicon and glass substrates. The silicon substrate is a 2-D rectangular array of pixel electrodes, and an active matrix circuit is embedded to control the electrical field injected to individual pixel electrodes. When the electric field of a particular pixel electrode is adjusted, the birefringence, and hence the phase retardation of the liquid crystal above the electrode, will change accordingly. A simplified structure of the LCoS SLM is shown in Figure 5.1.

One can easily infer that if the electrical field of the active matrix circuit is adjusted to make the phase-shifting of the 2-D array pixels in the SLM similar to that of a phase-only hologram, the light reflecting from it will reconstruct the 3-D image that is represented by the hologram.

Basically there are two modes of optical reconstruction of a hologram by an SLM. The first mode is holographic projection, whereby the diffracted optical waves emitted by the hologram are projected onto a screen to show the reconstructed image. The screen can be a plane that is parallel to the hologram, or a 3-D surface. If a divergent beam illuminates the hologram, the size of the projected image will increase linearly with the distance between the screen and the hologram. On the other hand, if a plane wave is used instead of a divergent beam, the size of the projected image will be similar in size to the SLM, independent of the location of the screen. In both cases, the hologram should be generated in a way that the reconstructed image is in focus on the entire screen. For a planar screen that is parallel to the hologram, the focused plane is set to be at the location of the screen.

In the second mode of optical reconstructed images, the hologram is displayed as a visible 3-D image on the SLM. These two modes of hologram reconstruction will be described in the subsequent part of this section. To enhance the quality of the reconstructed image, the size of a hologram should be large, and the resolution should be high. A large SLM with fine pixel size, therefore, is desirable for displaying a hologram. The size and resolution of an SLM can be expressed by a dimensionless term known as the space–bandwidth product (SBP). The SBP is the product of the display area and the resolution of an SLM. Suppose an SLM device comprises of M columns and N rows of pixels, each being a square cell of side length δ_d meters. The size and bandwidth of the SLM are $(M\delta_d \times N\delta_d)$ m^2, and $(\delta_d)^{-1}$ for both dimensions, resulting in SBP = $(M\delta_d \times N\delta_d)(\delta_d)^{-1}(\delta_d)^{-1}$ =$(M \times N)$.

5.2.2 Holographic Projection

A typical optical setup for holographic projection with a phase-only hologram and a phase-only LCoS SLM is shown in Figure 5.2(a). Assuming a planar screen that is parallel to the hologram, a phase-only hologram is generated from a given source image located at a distance z from the hologram plane. It is then displayed on a phase-only SLM. Next, a point laser beam is converted into an optical plane wavefront with a beam expander, and taken to illuminate the SLM through the beam splitter. Each pixel on the SLM, upon impinging by the incoming beam, scatters a wavefront that is modulated by a phase-shift equivalent to that of the hologram pixel it represents. A screen is placed at a distance z from the hologram to show the projected image, which is sometimes referred to as a "real image." The collection of optical wavefronts of all the pixel electrodes is reflected by the beam splitter, forming a focused optical reconstructed image on the screen. In this configuration, the illumination beam, the axial direction of the hologram, and the projected image are situated along the same optical path. As such, it is referred to as an inline holographic projection system.

A major disadvantage of the setup in Figure 5.2(a) is that the illumination beam and the optical waves emitted by the SLM are attenuated by the beam splitter. This will lead to a dim reconstructed image as a large proportion of the optical beam has been diverted to other locations. An alternate setup to overcome this problem is to adopt the slightly off-axis arrangement, as in Figure 5.2(b). The optical paths of the illumination beam and the holographic waves are separated. In this off-axis arrangement, the reconstructed image is brighter as the expanded laser beam is used to illuminate the hologram directly.

5.2.3 Holographic Display

A holographic projector is similar to a traditional projector; the main difference is that the focusing of the projected image is produced by the computation of the hologram rather than by using an optical lens. For holographic projection, the focusing surface on which the reconstructed image is formed is fixed by the geometry of the screen. As such, a holographic projector is limited to showing a 2-D image when it is projected onto a flat surface, rather than the full functionality of a hologram's true 3-D image. This limitation can be overcome with a 3-D holographic display, whereby the reconstructed image of a hologram is conveyed as a "virtual image," rather than a "real image." In a holographic projector, a real image is

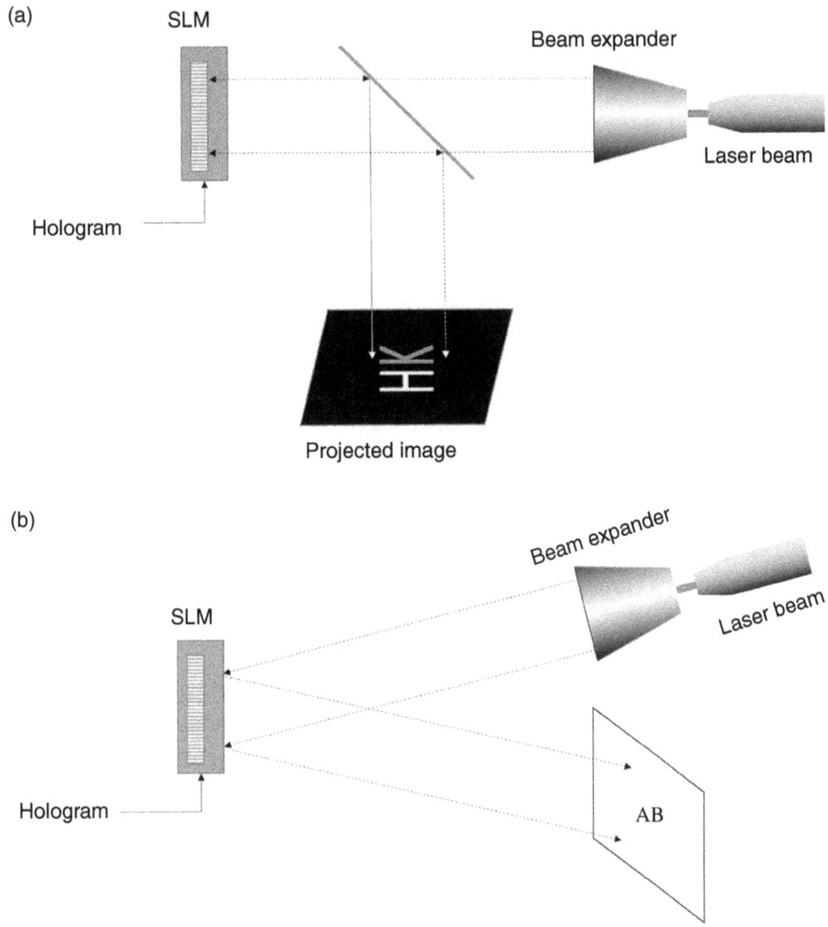

Figure 5.2 (a) An inline holographic projection unit. (b) An off-axis holographic projection unit.

projected from the hologram onto a screen, forming a visible image. A virtual image is a 3-D image that is formed behind the hologram. Observing a virtual image is like looking at a 3-D scene through a window, with the viewing zone limited by the coverage of the SLM. As an observer views through the window at different angles, different parts of the 3-D scene will become visible. In addition, when the viewpoint is shifted, the disparity between objects at different depths in the scene will also change, as if the observer is looking at a real physical environment.

5.3 Holographic Encryption

After learning the primary role of phase-only holograms in image projection and display, we shall now have a look at some practical applications of phase-only holograms in cryptography. Cryptography, also known as encryption, is the art of

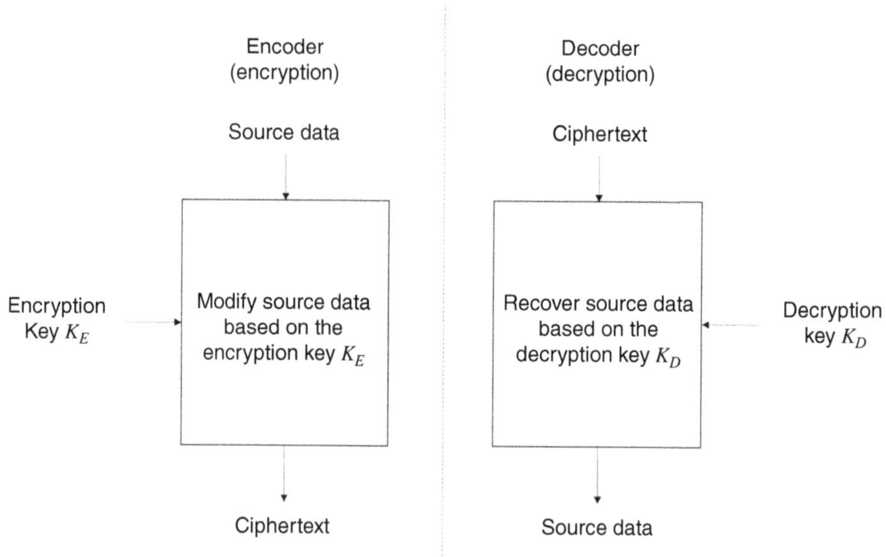

Figure 5.3 Fundamental framework of cryptography.

changing numerical data to a form that can be interpreted by someone with the knowledge of how to revert the changes. It is an important technology for protecting private information from illegitimate access. Modern internet technologies, such as cloud and optical networks, are fully capable of storing and delivering big data sets that are highly vulnerable to security threats and malicious attacks. The problem is particularly critical when the service provider is an unknown common third party, whereby data could be eavesdropped upon during the course of distribution. If a communication link is being eavesdropped upon, all the sensitive contents in the data stream will be exposed, which can lead to severe security and financial impacts. To prevent this, cryptography is mandatory for maintaining the confidentiality of sensitive data and preventing it from being deciphered by unauthorized parties. Data that has been encrypted is generally known as ciphertext. Over the years, numerous research attempts have been conducted in the area of cryptography, resulting in different forms of realization, and some of them are rather complicated. The basic concept of cryptography is illustrated and explained with the simple block diagrams in Figure 5.3.

Cryptography comprises two parts. The first part is known as the encoder, which converts the source data into a ciphertext using a secret encryption key K_E. The key can be in many forms, such as an image, a character string, or a mathematical operation on the input data. The goal is to generate a ciphertext that is significantly different from, and cannot be easily reverted back to, the source data. The second part of cryptography, known as the decoder, carries out the decryption process to revert the ciphertext to the source data with a decryption key K_D. Most encryption methods are symmetric, meaning that the encryption and decryption keys are identical (i.e., $K_D = K_E$). If the encryption key is exposed, the source data can be easily obtained from the ciphertext.

Many intruders, commonly known as cryptanalysts, have been attempting to deduce the secret keys of different encryptions through hacking attacks.

A well-known approach to deduce the encryption key in symmetric encryption is the "known plaintext attack" (KPA), with which the secret key (which is the same in both encryption and decryption) is deduced from the relation between a known source data set and its ciphertext. For example, a simple encryption that is performed by changing each value in the data set $S = [32, -200, 32, -128, -200, 32, 32, -128]$ into a different one, resulting in the ciphertext $S_e = [-16, 112, -16, 5, 112, -16, -16, 5]$. The encryption/decryption key is a mapping between the original values $S = [32, -200, 32, -128, -200, 32, 32, -128]$ with the modified values $[-16, 5, 112]$. Suppose the source data and the ciphertext are both available, the secret key can be easily derived by comparing each element in S with its corresponding counterpart in the ciphertext S_e.

Optical encryption is a variant of classical encryption, mainly applied in encrypting images. Instead of encrypting the source image directly, optical encryption methods are based on modifying the optical waves of a source image along its path of propagation. This approach is more secure than classical encryption, as the optical waves of a source image can be very different from the image itself. Hence even if the source image and its ciphertext are both available, it will be improbable to deduce the modifications that have been made to the optical waves emitted by the source image. Holographic encryption is a kind of optical encryption, whereby the hologram of the source data is encrypted. In this chapter we focus on holographic encryption of phase-only holograms. A hologram is basically used to represent pictorial information; to be explained later, it can also be employed to store and encrypt digital data.

5.3.1 Optical Cryptography

Most of the traditional encryption methods developed to date are designed to operate on the raw data directly. As mentioned before, such a kind of encryption can be cracked using KPA. In the past few decades, rapid advancements in optics have led to the emergence of optical encryption, adopting the theory of optics. Being different from data encryption in general, optical encryption is primarily used to encrypt optical images such as digital photographs. Despite the restricted scope of application, image encryption is important nowadays as there is a huge amount of image data transactions in contemporary social networks, wireless channels, and video broadcasting media. In optical encryption, the image to be encrypted is illuminated with a light source and the optical waves emitted from it are modified (based on an encryption key) at certain point(s) in the course of propagation toward a target destination. Grossly speaking, it is the light waves of the image, rather than the image itself, that are encrypted. Holographic encryption is a kind of optical encryption in which the wavefront selected for encryption is a hologram of the source image. A hologram is a 2-D complex image representing a 3-D image. Commonly, a hologram image comprises high-frequency fringe patterns that offer little clue on the image they represents. It is thus improbable to deduce the encryption key based on the ciphertext and the source image of a hologram.

5.3.2 Double Random Phase Optical Encryption

Effective optical/holographic encryption techniques developed to date are mostly based on the double random phase encoding (DRPE) framework. The method was first developed and reported by Philippe Refregier and Bahram Javidi in 1995 [1]. It is mainly used for encrypting optical images with which the light wave is encrypted in the spatial domain, and subsequently in the frequency space. The schematic of DRPE encryption is shown in Figure 5.4(a).

The source image $I(m,n)$ to be encrypted is illuminated with an on-axis coherent plane wave. The optical wave emitted by the source image is passed through a random phase mask $M_1(m,n) = e^{i\varphi_1(m,n)}$ ($\varphi_1(m,n)$ is a random quantity within the range $[0, 2\pi)$), which is taken to be the first encryption key. The light wave after passing through the random phase mask, given by $I_N(m,n) = I(m,n)e^{i\varphi_1(m,n)}$, is Fourier transformed to its frequency spectrum $\widetilde{I_N}(\omega_m, \omega_n)$ with lens L1. A random phase mask, $M_2(\omega_m, \omega_n) = e^{i\varphi_2(\omega_m,\omega_n)}$, taken as the second encryption key, is placed at the Fourier plane containing the spectrum of $I_N(m,n)$. Finally, the noise-added spectrum, denoted by $\widetilde{I_N}(\omega_m, \omega_n)e^{i\varphi_2(\omega_m,\omega_n)}$, is inverse transformed into the spatial domain with lens L2, resulting in a DRPE ciphertext $C(m,n)$. Denoting the forward and inverse Fourier transform operations as $\mathcal{F}$ and $\mathcal{F}^{-1}$, respectively, the ciphertext is given by

$$C(m,n) = \mathcal{F}^{-1}\{\mathcal{F}[I(m,n)M_1(m,n)]M_2(\omega_m, \omega_n)\}. \tag{5.1}$$

The decoding process is simply the reversal of the encryption, as shown in Figure 5.4(b). The ciphertext is illuminated with a coherent plane wave, and the optical wave emitted from it is converted into the frequency spectrum with lens L1. A phase mask $M_2^{\dagger}(\omega_m, \omega_n)$, which is the conjugate of $M_2(\omega_m, \omega_n)$, is placed on the Fourier plane. Upon exit from the phase mask, the light wave is inverse Fourier transformed to give a decrypted image $I_D(m,n)$ on the image plane. Mathematically:

$$\begin{aligned} I_D(m,n) &= \mathcal{F}^{-1}\{\mathcal{F}[C(m,n)]\mathrm{M}_2^{\dagger}(\omega_m, \omega_n)\} \\ &= I(m,n)M_1(m,n) = I(m,n)e^{i\varphi_1(m,n)}. \end{aligned} \tag{5.2}$$

With the use of a digital CCD camera, the magnitude of the decrypted image, given by $I(m,n)^2$, is recorded.

The DRPE has two major disadvantages. First, the optical setups of the encoder and the decoder are quite complicated, as part of the process has to be conducted in Fourier space. Second, according to the evaluation on the methods in [2–9], the DRPE is vulnerable to different kinds of attacks. In particular, if the encoder is available, the encryption keys can be deduced with the chosen plaintext attack (CPA) and/or the KPA. In these two kinds of attacks, the source images and the ciphertexts are both available, and the encryption key can be revealed by analyzing the relation between them. According to [9], the random phase mask $M_2(\omega_m, \omega_n)$ could be easily deduced from the ciphertext of a Dirac impulse (i.e., a single dot). There are different variations on the implementation of the DRPE method to increase its resistance to attacks, such

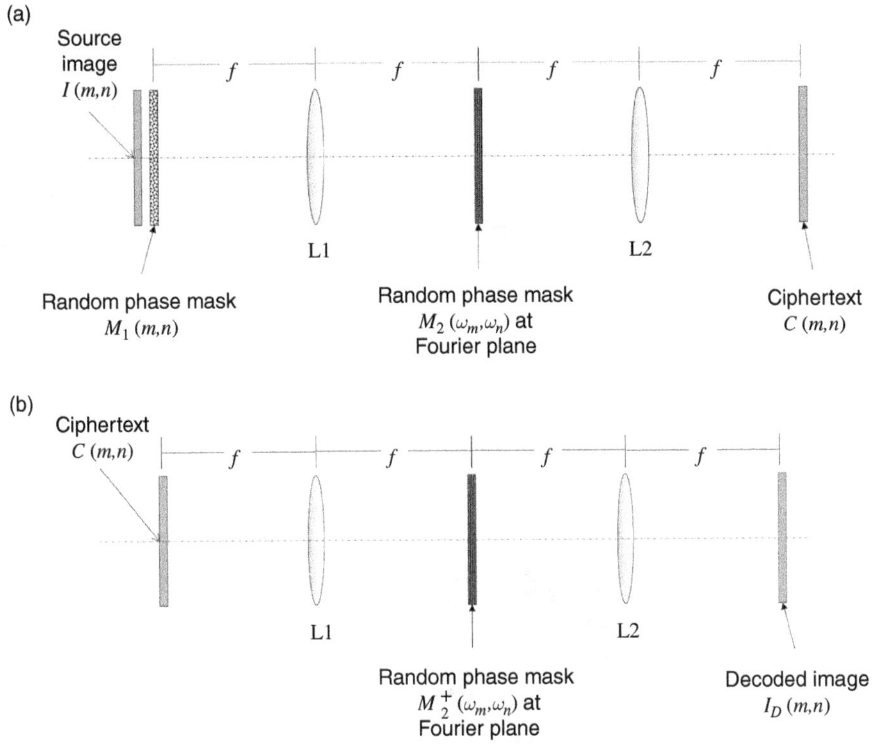

Figure 5.4 (a) Optical encryption with the DRPE framework. (b) Optical decryption with the DRPE framework.

as, but not limited to, phase reservation and compression [10], photon counting [11], space multiplexing of multiple images [12], fractional Fourier transform [13,14], randomized lens-phase function [15], and Arnold transform [16]. Apart from image encryption, DRPE has been extended to encrypt digital data by embedding it in a QR code [17,18].

5.3.3 Single Random Phase Holographic Encryption

Inspired by DRPE, a low-complexity holographic encryption method known as single random phase encryption (SRPE) was proposed in [19]. The concept of SRPE is shown in Figure 5.5(a). A 3-D source image is input to a stochastic hologram generator. In the generator, the source image is first modified in a random manner that retains its visual quality, and then placed in a larger global image $G(m,n)$. In the figure, the source image is scaled in size and added with a text message "ABC" that is not overlapping with the image. The modified image is then converted into a phase-only hologram $H_P(m,n)$. By adding a random phase mask $M(m,n)$ (the encryption key), the phase-only hologram is converted into a SRPE hologram $H_E(m,n) = H_P(m,n)M(m,n)$ (ciphertext). As is explained in more detail later, the process is equivalent to a one-time pad encryption generally considered to be an "unbreakable" cipher method.

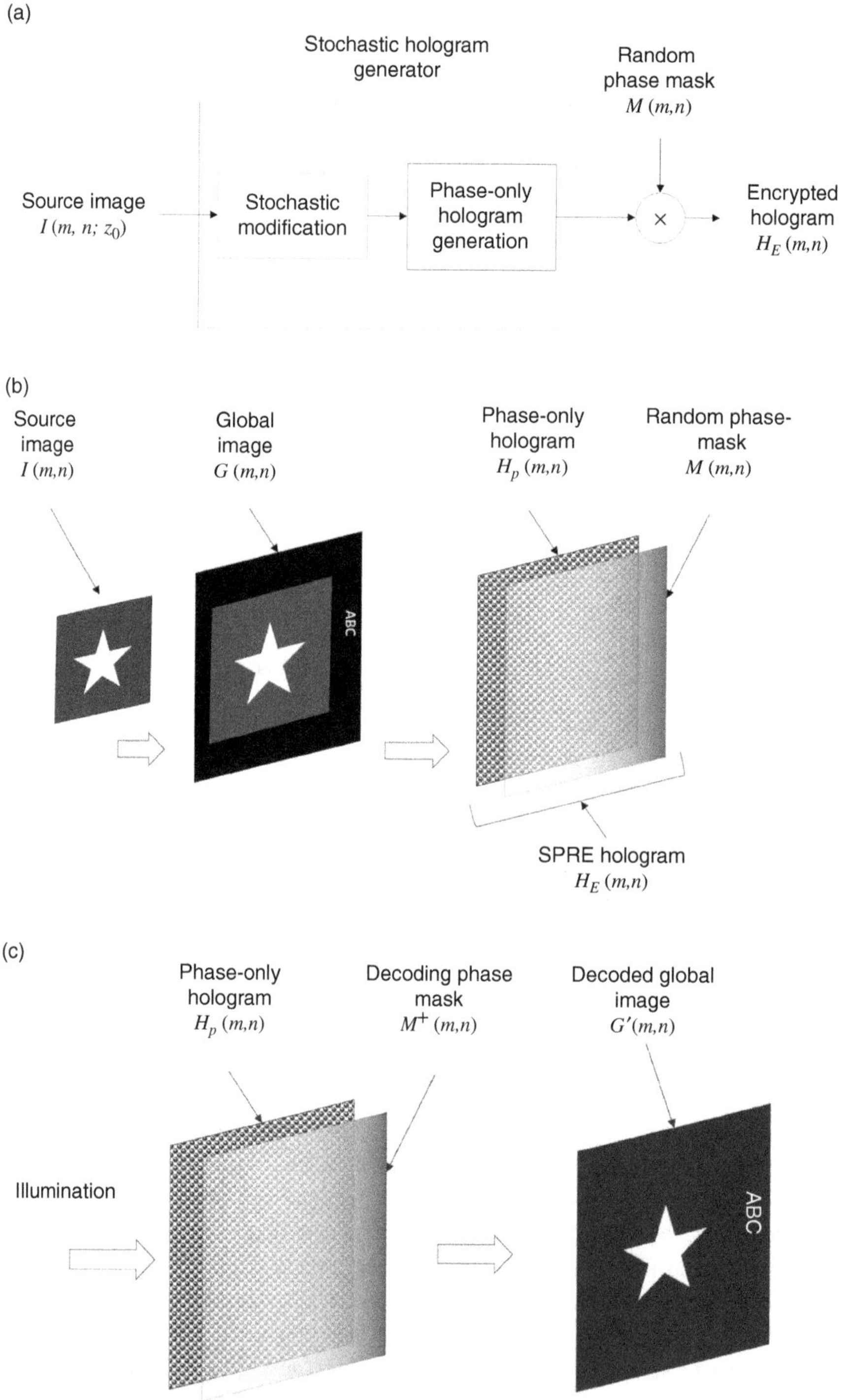

Figure 5.5 (a) Concept of the single random phase encryption (SRPE) method. (b) Implementation of the SRPE encoder. (c) Implementation of the SRPE decoder.

Implementation of the SRPE encoder and decoder is shown in Figures 5.5(b) and 5.5(c), respectively. In the encoder, the source image is embedded into a global image and converted into a phase-only hologram $H_P(m,n)$. The latter is them overlaid with a random phase mask $M(m,n)$, resulting in an SRPE hologram $H_E(m,n)$. In the decoder, the encrypted hologram $H_E(m,n)$ is overlaid with a phase mask $M^{\dagger}(m,n)$, which is the complex conjugate of $M(m,n)$, thus recovering the phase-only hologram $H_P(m,n)$ of the global image. When $H_P(m,n)$ is illuminated with a coherent beam, the global image and the modified source image that is embedded in it will be reconstructed to form a visible image. Figures 5.5(b) and 5.5(c) show that the complexity of the SRPE method is substantially lower than the DRPE. However, since the ciphertext is representing the global image rather than the source image, it will be improbable to apply KPA or CPA to deduce the encryption key.

The SRPE method is simple in structure and has the following advantages:

1. The source image is modified in an unknown and random manner, so that encryption of the actual image is unknown. The absence of knowledge on what is represented by the phase-only hologram and its encrypted data makes it difficult to deduce the encryption key by plaintext attacks.
2. For the same input source image, a vastly different phase-only hologram results in each round of encryption. Hence, it is not possible to deduce the encryption key by analyzing the correlation (or difference) between a large set of similar source images and their corresponding ciphertexts.
3. After adding the random phase mask, which is taken as the encryption key, it is even more difficult to deduce the encryption key through analyzing the relation between the input image and the SRPE hologram.
4. Upon decryption, the perceptual quality of the original source image is chiefly preserved favorably even though a certain part of it has been distorted.

Following is an explanation of the detail formulation of the SRPE method, which can be divided into three stages.

Stage 1: Stochastic Modification of the Source Image

The source image is modified randomly in a manner that is unknown to the person encrypting the image. Assume the encryption device, whether it is implemented with software or hardware, is a black box so that the details within it are concealed from the operator. To avoid causing too much change to the source image, the random modification is limited to a moderate amount of geometrical transformation (such as scaling and translation). After modifying the source image $I(m,n)$, it will be pasted into an arbitrary position in a larger global image $G(m,n)$. The rest of the areas in the global image that are not occupied by the source image will be filled with randomly generated content $P(m,n)$. Mathematically, the global image is given by

$$G(m,n) = T[I(m,n)] \bigcup P(m,n), \tag{5.3}$$

where $T[\ ]$ represents some form of geometrical transform. Equation (5.3) constitutes a stochastic modification of the source image, so that the global image is substantially different from the original source image. The global images generated in different runs of the encryption process will also deviate from one another. This mechanism ensures that neither the cryptanalyst nor the one who performs the encryption will have knowledge of what is being encrypted. Without information on the source data, it will be improbable, if not impossible, to draw any relationship between the ciphertext and the source image. By the same argument, the stochastic modification will also prevent the encryption key from being exposed through large-scale CPA, as the images being utilized to probe the encryption engine will also change in an unknown manner prior to their encryption.

Stage 2: Converting the Global Image to a Phase-Only Hologram

The global image is first converted to a digital Fresnel hologram by convolving the global image with free space impulse response of the light propagation $h(m,n;z_0)$ as

$$H(m,n) = G(m,n) * h(m,n;z_0). \tag{5.4}$$

The free space impulse response of light propagation is given by $h(m,n;z_0) = \frac{i}{\lambda z_0} exp\left\{\frac{i2\pi}{\lambda}\sqrt{(m\delta_d)^2 + (n\delta_d)^2 + z_0^2}\right\}$, where λ is the wavelength of light, and δ_d is the pixel size of the hologram. As the hologram is generated from the global image, it is referred to as the global hologram.

The complex-valued hologram $H(m,n)$ is then converted into a global phase-only hologram $H_P(m,n) = exp[i\theta(m,n)]$ with either uni-directional or bi-directional error diffusion.

Stage 3: Encrypting the Phase-Only Hologram with a Random Phase Mask

In this final stage, a random phase mask $M(m,n) = exp[i\varphi(m,n)]$ (where $\varphi(m,n)$ is a random value in the range $[0,2\pi)$) is overlaid onto the global phase-only hologram, resulting in an encrypted hologram

$$H_E(m,n) = H_P(m,n)M(m,n) = exp\{i[\theta(m,n) + \varphi(m,n)]\}. \tag{5.5}$$

In the decoder, as shown in Figure 5.5(c), the global phase-only hologram is recovered by overlaying the conjugate of the random phase mask $M(m,n)$ onto $H_E(m,n)$:

$$\begin{aligned} H_P(m,n) &= H_E(m,n)M^\dagger(m,n) \\ &= [H_P(m,n)M(m,n)]M^\dagger(m,n) = H_P(m,n). \end{aligned} \tag{5.6}$$

The random phase mask $M(m,n)$ can be taken as the secret key for encrypting and decrypting the global image, which is different from the source image $I(m,n)$. Hence, given $I(m,n)$ and $H_E(m,n)$, it is not possible to deduce the secret key $M(m,n)$. After

Figure 5.6 (a) Source image "Fish1." (b) Global image. (c) SRPE hologram of the global image. (d) Reconstructed image of the decoded SRPE hologram on the focused plane.

decryption, the global image can be reconstructed from the hologram through optical or numerical reconstruction. From the reconstructed global image, the source image, which is separated from the randomly added contents, will appear with good visual quality.

Numerical simulation of the SRPE method is conducted with the source code in Section 5.5.1. The source image "Fish_1" in Figure 5.6(a) is embedded into the larger canvas "Canvas" comprising of a QR code and a text message "Single random phase encryption," resulting in a global image as shown in Figure 5.6(b). A phase-only hologram $H_P(m,n)$ is generated from the global image with uni-directional error diffusion and infused with random phase noise $M(m,n)$ to give an SRPE hologram $H_E(m,n)$, as shown in Figures 5.6(c). The SRPE hologram is decoded by being overlaid with the conjugate of the random phase mask $M^{\dagger}(m,n)$, and the reconstructed image is shown in Figure 5.6(d). It can be seen that apart from some mild noise contamination, the reconstructed image is similar to the original one.

5.3.4 Enhanced Single Random Phase Holographic Encryption

Despite the effectiveness of SRPE, the source image is modified after encryption. Although the modification does not affect the visual quality, it may not be acceptable in some applications. In view of this, enhanced single random phase encryption (ESRPE) was proposed in [20]. In comparison with SPRE, the ESRPE has similar advantages, but will not lead to any alteration to the source image.

The ESRPE method comprises four stages, as shown in Figure 5.7 and outlined here.

In the first stage, the source image $I(m,n)$ to be encrypted is converted into an off-axis complex hologram based on angles of incidence θ_m (along the horizontal direction) and θ_n (along the vertical direction). If $\theta_m = 0$ and $\theta_n = 0$, the complex hologram is equivalent to an on-axis or inline hologram. The source image can be a planar image or a 3-D image. When the off-axis hologram of the source image is illuminated with a plane wave, the source image will be reconstructed on the focused plane oriented at angles θ_m and θ_n along the horizontal and vertical directions, respectively, from the normal of the hologram plane. The hologram being generated from the source image is known as the source hologram.

In the second stage, one or more secondary images $I_k(m,n)|_{1\le k\le K}$ (where $K \ge 1$ is the total number of secondary images) are randomly generated or selected from a pool of images that is unknown to the operator, and are converted into an off-axis complex hologram. For simplicity of illustration, only one secondary image $I_1(m,n)$ is shown in Figure 5.7. Each off-axis hologram is generated based on a reference beam at an angle of incidence $\theta_{k;m}$ (along the horizontal direction) and $\theta_{k;n}$ (along the vertical direction).

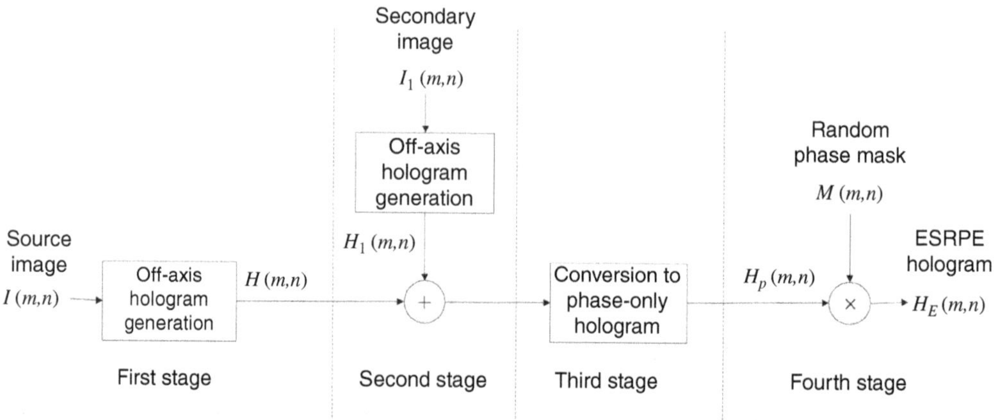

Figure 5.7 The enhanced single random phase encryption (ESRPE) method.

When the off-axis hologram of the secondary image is illuminated with a plane wave, the secondary image will be reconstructed on the focused plane oriented at angles $\theta_{k;m}$ and $\theta_{k;n}$ along the horizontal and vertical directions, respectively, from the normal hologram plane. The hologram being generated from a secondary image is known as the secondary hologram. The same applies if there is more than one secondary image, in which case there will be a set of secondary images with corresponding secondary holograms.

In the third stage, the source hologram and the secondary hologram(s) are summed up into a global hologram $H_G(m,n)$ and converted into a global phase-only hologram $H_P(m,n)$ with existing techniques such as, but not limited to, error diffusion [21,22]. The global phase-only hologram is very different from the source hologram and any of the secondary hologram(s). As the secondary image is randomly generated or selected from unknown sources, even for identical source images, a totally different global phase-only hologram will be generated in each run of the encryption process.

In the fourth stage, a random phase mask $M(m,n) = e^{i\varphi(m,n)}$ is added to the global phase-only hologram $H_P(m,n)$, resulting in an ESRPE hologram $H_E(m,n)$. The latter is a ciphertext hologram that bears little correlation to the source image or the secondary image(s).

The mixing of the holograms of the source image and one or more secondary images ensures no prior knowledge of what is being encrypted, so that there is no direct relation between the ESRPE hologram and the source image. It also increases the difficulty of deducing the encryption key through large-scale CPA (e.g., encrypting a large number of source images, each with a small amount of incremental change, to locate the affected regions of the ciphertext hologram).

In the decoder, the ESRPE hologram is overlaid with the conjugate of the random phase mask (encryption key) and illuminated with a coherent beam. The source and the secondary images will be reconstructed at the focused plane along orientations $H_E(m,n)$ and $(\theta_{k;m}, \theta_{k;n})$, respectively. If the angular separation between (θ_m, θ_n) and the orientation(s) of the secondary holograms $(\theta_{k;m}, \theta_{k;n})$ is large enough, the images will be formed at non-overlapping areas on the focused plane, and the source image can be observed visually at its original position without any geometrical change. The specific formulation of each stage of the proposed method is described in the following sub-sections.

Stage 1: Converting the Source Image into an Off-Axis Hologram

For simplicity of explanation, the source is assumed to be a 2-D planar image $I(m,n)$, which is parallel to and at an axial distance z_0 from the hologram. The source image is first converted into an off-axis complex Fresnel hologram $H(m,n)$; $R(m,n)$ is an inclined plane wave with an angle of incidence (θ_m, θ_n).

$$H(m,n) = R(m,n) \times [I(m,n) * h(m,n;z_0)], \tag{5.7}$$

where $R(m,n)$ is an inclined plane wave with an angle of incidence (θ_m, θ_n).

Stage 2: Converting Secondary Images into an Off-Axis Hologram

The secondary images are assumed to be 2-D planar images $I_k(m,n)|_{1\leq k\leq K}$, which are parallel to and at an axial distance z_k from the hologram. However, more secondary images can be added by the same principle. Similar to stage 1, an off-axis hologram $H_k(m,n)$ is generated for the kth secondary image with the angle of incidence of the plane wave $R_k(m,n)$ at angles of incidence $(\theta_{k;m},\theta_{k;n})$.

Stage 3: Generating a Global Hologram

In this stage, the source and the secondary holograms are integrated into a new hologram $H_G(m,n)$:

$$H_G(m,n) = H(m,n) + \sum_{k=1}^{K} H_k(m,n). \tag{5.8}$$

Subsequently, $H_G(m,n)$ is converted into a global phase-only hologram $H_P(m,n)$ with the error diffusion method. In $H_P(m,n)$, the magnitude of all the pixels is set to a constant value.

Stage 4: Converting the Global Hologram into a ESRPE Hologram

Finally, the global hologram is multiplied with a random phase mask $M(m,n) = e^{i\varphi(m,n)}$, resulting in an ESRPE hologram given by

$$H_E(m,n) = H_P(m,n)e^{i\varphi(m,n)}. \tag{5.9}$$

The ESRPE hologram is taken as the encrypted (ciphertext) hologram, with the phase mask $e^{i\varphi(m,n)}$ as the private encryption key. When the ciphertext hologram is overlaid with the conjugate of the phase mask, and illuminated with a coherent beam, the global image, comprising the source image and the secondary image(s), will be reconstructed. The decryption process can be conducted through optical or numerical reconstruction as

$$H_P(m,n) = H_E(m,n)e^{-i\varphi(m,n)}. \tag{5.10}$$

Numerical simulation on the ESRPE method is conducted with the MATLAB code in Section 5.5.2. The source image "Fish_2" and a secondary image "Flower_1" are shown in Figures 5.8(a) and 5.8(b), respectively. Both images are located at an axial distance of 0.3 m from the hologram plane. An inline hologram is generated for the source image, and an off-axis hologram is generated for the secondary image with $(\theta_{k;m},\theta_{k;n}) = (0.8^\circ, 0^\circ)$. A global hologram is obtained by summing up the holograms of the source and the secondary images. The global image is then converted into a phase-only hologram with uni-directional error diffusion, as shown in Figure 5.8(c), and mixed with a random phase mask $M(m,n)$ into an ESRPE hologram $H_E(m,n)$ in Figure 5.8(d). Subsequently, the ESRPE hologram is decoded by integrating the conjugate of the random phase mask and the reconstructed image of the decoded

Figure 5.8 (a) Source image "Fish_1." (b) Secondary image "Flower_1." (c) Global phase-only hologram representing the source and the secondary images. (d) ESRPE hologram. (e) Reconstructed image of the decoded ESRPE hologram on the focused plane. (f) Enlarged view of the reconstructed source image.

hologram $H_P(m, n)$ on the focused plane, as shown in Figure 5.8(e). The reconstruction of the source and the secondary images are separated from each other, and yet the reconstructed source image is similar to the original one, as illustrated in Figure 5.8(f).

5.3.5 Multiple-Image Holographic Encryption with Arnold Transform

In SRPE and ESRPE, an encrypted image can be decrypted and reconstructed optically as a visible image through overlaying the conjugate of the phase mask on the encrypted hologram. These two methods are only effective in encrypting a single source image. More sophisticated encryption of multiple images in a single phase-only hologram can be accomplished if part of the encoding and decoding process can be conducted numerically. Recently, a method has been proposed in [23] for encrypting multiple images in a single phase-only hologram by incorporating the Arnold transform. Briefly, Arnold transform is a reversible chaotic mapping of a torus into itself. The Arnold transformed data set can be fully recovered through an inverse process. This technique can be widely applied in scrambling, encryption, and data hiding [24–26]. Referring back to the method in [23], a set of source images are each converted into a hologram with the grid-cross downsampling (GCD) method, and transformed with

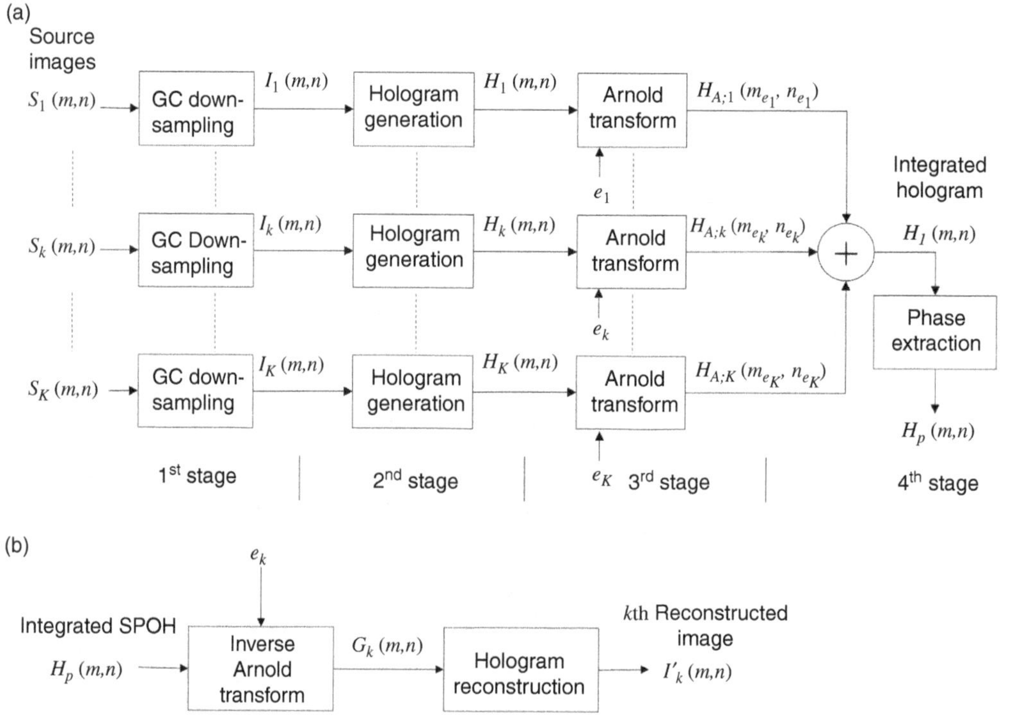

Figure 5.9 (a) Encrypting multiple images into a single integrated sampled phase-only hologram (ISPOH). (b) Decrypting the *k*th image from the ISPOH.

the Arnold transform with a unique number of iterations. The Arnold transformed holograms are summed up into a single integrated hologram, and the phase component is retained as an integrated sampled phase-only hologram (ISPOH) that encrypts all the source images. To recover a particular image, the integrated phase-only hologram is decrypted by applying the inverse Arnold transform with the corresponding number of iterations. As the holograms of the source images are generated with the GCD method, the decrypted image can be observed by simply illuminating the phase-only hologram with a coherent beam.

The encryption and decryption process is as follows. The encryption process comprises three stages, as illustrated in Figure 5.9(a); the decryption process is shown in Figure 5.9(b). First we shall study the encryption process. A set of K images, denoted by $S_k(m,n)|_{1\leq k\leq K}$, is to be encrypted into a phase-only hologram $H_P(m,n)$. For clarity and without loss of generality, the source images are assumed to be planar, parallel to the hologram, and located at an identical axial distance of z_0 from the hologram.

Stage 1: Generating Digital Fresnel Holograms With the Grid-Cross Downsampling (GCD) Method

One of the major objectives of the method is to encrypt the source images into a phase-only hologram, in order to display an image of reasonable visual quality with the decrypted hologram on a single phase-only SLM. To achieve this goal, the downsampling method is employed to generate the hologram of each source image. As the GCD method has been described in detail in Chapter 3, only a brief outline is given here. In the GCD method, the intensity of each image is downsampled with a grid-cross lattice prior to the conversion to a hologram.

Let τ and % denote the downsampling interval and the modulus operator, respectively, and the grid-cross downsampling lattice $L(m,n)$ is,

$$L(m,n) = \begin{cases} 1 & (pq = 0) \text{ or } (p = q) \text{ or } [p = (\tau - 1) - q] \\ 0 & \text{otherwise} \end{cases}, \tag{5.11}$$

where $p = (m\%\tau)$ and $q = (n\%\tau)$.

After downsampling the grid-cross lattice, the downsampled source images are given by

$$I_k(m,n)|_{1 \le k \le K} = S_k(m,n)L(m,n). \tag{5.12}$$

Next, each downsampled source image $I_k(m,n)$ is converted into a Fresnel hologram $H_k(m,n)$ by convolving with the free space spatial impulse response of propagation of light $h(m,n)$, resulting in a digital Fresnel hologram mathematically as

$$H_k(m,n)|_{1 \le k < K} = I_k(m,n) * h(m,n;z_0), \tag{5.13}$$

where $*$ is the 2-D convolution operator and $h(m,n;z_0)$ is the free space impulse response of light propagation.

Stage 2: Multiple Arnold Transform of Holograms

In the first stage, holograms of different source images are generated with Fresnel diffraction, to be combined into a single integrated hologram. However, the holograms cannot be directly summed up into an independent entity, as their contents will be merged and not separable in the future. Hence, in the second stage each of the holograms generated in the first stage is encrypted and merged with each other. The encryption is conducted with repetitive rounds of Arnold transforms (which is referred to as multiple Arnold transform) so that each individual hologram can be extracted in the decryption process. As reported in [27], the Arnold transform is a scrambling technique whereby the locations of pixels are shuffled through a 2-D transformation matrix. In each round of transformation, the position of each pixel of the current hologram is moved to a new position, resulting in a new hologram. In order to enable individual holograms to be extracted from the integrated hologram, it is mandatory that the number of rounds of Arnold transform should be different for each hologram. Consider the kth hologram $H_k(m,n)$ of size $N \times N$. Let $H_{A;k}(m_j,n_j)$ and $H_{A;k}(m_{j-1},n_{j-1})$ denotes the hologram after j and $j-1$ rounds of Arnold transforms, respectively. A pixel at location (m_j,n_j) in $H_{A;k}(m_j,n_j)$ is copied from the pixel at location (m_{j-1},n_{j-1}) of $H_{A;k1}(m_{j-1},n_{j-1})$ according to the following equation:

$$\begin{bmatrix} m_j \\ n_j \end{bmatrix} = \begin{bmatrix} 1 & a \\ b & ab+1 \end{bmatrix} \begin{bmatrix} m_{j-1} \\ n_{j-1} \end{bmatrix} \%N, \tag{5.14}$$

where $A = \begin{bmatrix} 1 & a \\ b & ab+1 \end{bmatrix}$ is the Arnold transform matrix, and $(a,b) = (2,1)$ are the transform parameters.

The pixel locations in an Arnold transformed image can resume their original positions with an inverse Arnold transform, given by

$$A^{-1} = \begin{bmatrix} ab+1 & -a \\ -b & 1 \end{bmatrix}. \tag{5.15}$$

Applying the inverse Arnold transform to the hologram $H_{A;k}(m_j, n_j)$, the pixels of hologram $H_{A;k}(m_{j-1}, n_{j-1})$ can be deduced from the following expression:

$$\begin{bmatrix} m_{j-1} \\ n_{j-1} \end{bmatrix} = A^{-1} \begin{bmatrix} m_j \\ n_j \end{bmatrix} \%N. \tag{5.16}$$

As an example, a pixel at $(m_{j-1}, n_{j-1}) = (100, 50)$ at the $(j-1)$th iteration with $N = 128$ will be relocated to a new position at $(m_j, n_j) = (72, 122)$ after applying a single round of Arnold transform. The pixel can be returned to its original position with the inverse Arnold transform.

Let e_k denote the number of Arnold transforms that have been applied to the kth hologram. The transformed hologram is represented by $H_{A;k}(m_{e_k}, n_{e_k})$. In order to retain the individuality of each hologram in the integrated hologram, the number of transforms applied to each hologram has to be different: $e_A = e_B \Leftrightarrow A = B$.

Stage 3: Generating the ISPOH

In this final stage, the Arnold transformed holograms are integrated into a single integrated hologram given by

$$H_I(m,n) = \sum_{k=1}^{K} H_{A;k}(m_{e_k}, n_{e_k}). \tag{5.17}$$

Subsequently, the phase component of the integrated hologram is retained as the ISPOH $H_P(m,n)$:

$$H_P(m,n) = \arg[H_I(m,n)], \tag{5.18}$$

where arg[] denotes the argument – that is, the phase component of the bracketed complex quantity. The value e_k, which corresponds to the number of Arnold transforms on the kth hologram, will be taken as the unique key parameter for retrieving the kth hologram from the ISPOH.

The decoding process is shown in Figure 5.9(b). To retrieve the kth hologram from the ISPOH, e_k rounds of inverse Arnold transforms are applied to the ISPOH

$H_P(m,n)$, resulting in a decoded phase-only hologram $G_k(m,n)$. Among all the holograms that have been encrypted in the $G_k(m,n)$, only the one corresponding to $H_{A;k}(m_{e_k}, n_{e_k})$ can be reverted to an approximation of $H_k(m,n)$, denoting this hologram by $H_k'(m,n)$. This is mixed with the rest of the holograms, which are not inverse transformed with the correct number of rounds of inverse Arnold transform. When the decoded hologram $G_k(m,n)$ is illuminated with a coherent beam, $H_k'(m,n)$ will be reconstructed, with a certain amount of distortion caused by removal of the magnitude component, to an image $I_k'(m,n)$ that is an approximation of the downsampled source image $I_k(m,n)$. The rest of the components in the ISPOH are reconstructed as noise data. Alternatively, $I_k'(m,n)$ can also be derived numerically through convolving $k(m,n)$ with $h^\dagger(m,n;z_0)$, the conjugate of the free space impulse response, as given by

$$I_k'(m,n) = G_k(m,n) * h^\dagger(m,n;z_0). \tag{5.19}$$

To further reduce the noise contamination, as well as avoiding the zero-order beam in optical reconstruction, the decoded hologram $G_k(m,n)$ can be converted into an off-axis hologram $G_k^a(m,n)$ by multiplying it with an inclined plane wave $R_{\theta_r}(m,n)$ as

$$G_k^a(m,n) = G_k(m,n)\ R_{\theta_r}(m,n), \tag{5.20}$$

where $R_{\theta_r}(m,n) = exp\left(\frac{i2\pi m}{\lambda} sin\theta_r\right)$.

The simulation on the multiple-image holographic encryption method is given in Section 5.5.3. Four source images $S_j|_{1\leq j\leq 4}$, as shown in Figure 5.10(a–d), are employed to illustrate the method. Referring to the schematic in Figure 5.9(a), each image is downsampled with the grid-cross lattice with the factor $\tau = 12$, and converted into a 1024×1024 Fresnel hologram. The axial distance between the image and the hologram planes is 0.3 m for all the source images.

To combine the four holograms into a single one, each hologram is transformed with e_j rounds of the Arnold transform. The value of e_j is set to $10 \times j$ so that each hologram is transformed with a unique round of Arnold transforms. This will permit holograms to be separable from each other in the future. Next, the transformed holograms are integrated into a single hologram $H_I(m,n)$, and an ISPOH $H_{ph}(m,n)$ is obtained by forcing the magnitude of all the pixels in $H_I(m,n)$ to unity.

In the decoding stage, the holographic information corresponding to the hologram of the jth image is recovered from $H_{ph}(m,n)$ with e_j rounds of inverse Arnold transforms, resulting in the phase-only hologram $G_j(m,n)$. The numerical reconstructed images of the four holograms from $G_j(m,n)|_{0\leq j\leq 3}$ are shown in Figure 5.10 (e–h). Apart from the degradations being caused by the grid-cross downsampling of the source images and the removal of the magnitude component in the integrated hologram, the numerically reconstructed images are similar to the source images.

Figure 5.10 (a–d) Source images $S_1 - S_4$, respectively. (e–h) Numerically reconstructed images $I_1' - I_4'$ from the ISPOH.

5.4 Holographic Steganography

Steganography is the art of concealing secret information, such as a message or an image, in another data set. However, since the host data set is modified after embedding the secret data, it is essential that the introduced changes do not cause observable differences. The embedding capacity, that is the amount of information that can be embedded into the host, is generally rather limited. Holographic steganography is embedding information in a hologram. As holograms are resistant to damage and capable of representing multiple images, the embedding capacity is significantly higher than that of an optical image such as a photograph. A number of methods for concealing information in a digital hologram are introduced in this section.

5.4.1 Data Embedded Error Diffusion Hologram

In Chapter 4 we saw that error diffusion is an effective means of converting a hologram into a phase-only hologram. Error diffusion is a local error compensation mechanism with which the error resulting from approximating a pixel in a certain way (e.g., converting it into a binary value) is distributed to other pixels in its vicinity. In converting a complex-valued hologram to a phase-only hologram with error diffusion, the magnitude of each hologram pixel is removed, and the error is added to the neighboring pixels. It has been proven that an error diffusion hologram is capable of preserving the source image with high quality, although only the phase component of the hologram is present. On this basis, a method was proposed in [28] for adding external data into a hologram, whereby the error resulting from the contamination and that arising from the removal of the magnitude component are combined and compensated with error diffusion. A phase-only hologram generated with this method is referred to as a data embedded error diffusion (DEED) hologram. Experimental results have demonstrated that a DEED hologram is capable of embedding an image that is of similar size to the hologram, and yet preserving favorable quality on both the hologram and the embedded image.

Generation of a DEED hologram is divided into three stages, as shown in Figure 5.11.

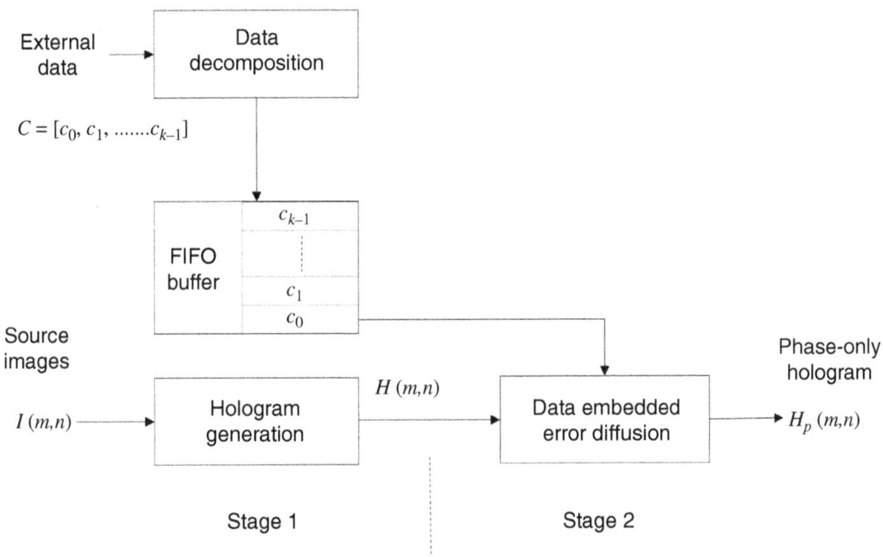

Figure 5.11 Schematic of the DEED hologram generation method.

Stage 1: Hologram Generation and Embedded Data Decomposition

In this stage, a digital Fresnel hologram $H(m,n)$ is generated from a source image with Fresnel diffraction. For the sake of simplicity, we have assumed that the source image is planar and parallel to the hologram plane. The magnitude and phase components of $H(m,n)$, denoted by $|H(m,n)|$ and $\angle\ [H(m,n)]$ (where $\angle[\ ]$ denotes the phase angle of the complex variable within the square brackets), are normalized into the range $[0,1]$ and $[0,255)$, respectively. As such, the phase angle of the hologram pixel is represented with eight bits. The embedded data is decomposed into a sequence of short chunks of data $C = [c_0, c_1, \ldots\ ..c_{K-1}]$, each chunk being represented with b_L bits, where $b_L < 8$, and queued into a first-in-first-out (FIFO) buffer. In the FIFO buffer, c_0 is the first member (head) and c_{K-1} is the last member (tail) of the queue.

Stage 2: Data Embedded Error Diffusion

In this stage, the complex-valued hologram $H(m,n)$ is added to external data in a pixel-by-pixel manner and converted into a phase-only hologram $H_P(m,n)$ with uni-directional error diffusion. The hologram is scanned pixel by pixel from the first row to the last row, and from the leftmost column to the rightmost column in each row. The magnitude of each hologram pixel is set to unity and the phase component is retained, converting it into a phase-only hologram pixel. The latter is then split into a lower-ordered byte of b_L bits and a higher-ordered byte of $b_H = (8 - b_L)$ bits. The head of the queue in the FIFO buffer is taken to replace the lower-ordered byte of the phase component. Subsequently, the head is removed from the FIFO buffer and the next member becomes the head of the queue. Suppose the current hologram pixel is located at the y_j row and the x_i column, and the current head of the queue is c_k. The current hologram pixel will be converted into a phase-only value:

$$H_P(m_i, n_j) = \{\angle\ [H(m_i, n_j)] \wedge b_M\} \vee c_k, \tag{5.21}$$

where b_M is a binary string comprising b_L bits of 0. An example is shown in Figure 5.12 with $\angle\ [H(m_i,n_j)] = [11000101]$, b_L=4, and $c_k = [1100]$. The revised pixel value is given by $H_P(m_i,n_j) = [11001100]$.

The error resulting from converting the current hologram pixel into $H_P(m_i,n_j)$ is diffused to its four unvisited neighboring pixels.

The source code for generating the DEED hologram is shown in Section 5.5.4. In the simulation, a Fresnel hologram $H(m,n)$ of the source image "Flower 2" in Figure 5.13(a) is generated. Next, uni-directional error diffusion is applied to convert the hologram into a phase-only hologram, and at the same time embedding it with a 2-D image "Grey," shown in Figure 5.13(b). The reconstructed image of the DEED hologram and the embedded image corresponding to $b_L = 2$ and 6 are shown in Figure 5.13(c. d) and 5.13(e, f), respectively.

From the results shown in Figure 5.13(c–f), it can be seen that the quality of the reconstructed images improves with a lower number of embedded bits (i.e., a smaller value of b_L), but the other way round for the embedded image.

$\angle[H(m_i, n_j)]$	1	1	0	0	0	1	0	1
b_M					0	0	0	0
$\{\angle[H(m_i, n)]\wedge b_M\}$	1	1	0	0	0	0	0	0
c_k					1	1	0	0
$H_P(m_i, n_j)$	1	1	0	0	1	1	0	0

Figure 5.12 Example showing embedding an external chunk of data into a phase-only hologram pixel.

Figure 5.13 (a) Source image "Flower_2." (b) Embedded image "Grey." (c) Reconstructed image of DEED hologram embedded with $b_L = 2$ bits. (d) Reconstructed image of DEED hologram embedded with $b_L = 6$ bits. (e) Image embedded in DEED hologram with $b_L = 2$ bits. (f) Image embedded in DEED hologram with $b_L = 6$ bits.

5.4.2 Image Embedded Error Diffusion Hologram

Section 5.4.1 shows that the DEED hologram is capable of embedding a certain number of data bits into each hologram pixel, at the expense of degrading the reconstructed image of the hologram. It is desirable to limit the bit length of the embedded data in each hologram pixel to fewer bits in order to preserve higher fidelity of the reconstructed image. The examples in Figure 5.13 reflect that if the embedded data is an image, its quality is rather poor if each of its pixels is truncated into 2 bits. This dilemma can be overcome if the data size of the image is reduced prior to embedding into a hologram, a process that is generally referred to as image compression. Attempts to address this issue have been reported in [29,30] based on block truncation coding and error diffusion. The following subsection explores these two methods and their performances, in terms of compression and fidelity, by reducing the data size of an image.

5.4.2.1 Block Truncation Coding

Block truncation coding (BTC) is a low-complexity image compression method that was developed by E. Delp and O. Mitchell in 1979 [31]. An image to be compressed is uniformly partitioned into non-overlapping square image blocks, and each block is compressed individually. The amount of computation is small, which is suitable for implementation with low-speed processors or hardware circuits. In general, an 8-bit gray level image can be compressed by four times with the BTC and still maintain high fidelity (in terms of PSNR) of over 30 db. As image blocks are compressed separately, the encoded data is highly resistant to damage and noise contamination. This implies that if a certain part of the compressed data is modified or lost, it will only affect the corresponding regions in the decrypted image, while the rest of the areas will remain intact. BTC can be divided into three stages as follows.

Stage 1: Block Partitioning

An input image $I(m, n)$ is uniformly partitioned into non-overlapping square blocks of size $\tau \times \tau$. For an image with horizontal and vertical extents of M and N, respectively, the total number of image blocks in an image is MN/τ^2, each composed of τ^2 pixels.

Stage 2: Mean and Standard Deviation Computation

In this stage, the mean and standard deviation of each image block is determined. Let b_j denote the jth image block in the image. The mean μ_j and standard deviation σ_j of the jth image block are:

$$\mu_j = \frac{1}{\tau^2} \sum_{p=0}^{\tau-1} \sum_{q=0}^{\tau-1} b_j(p, q), \tag{5.22}$$

and

$$\sigma_j = \sqrt{\frac{\sum_{p=0}^{\tau-1}\sum_{q=0}^{\tau-1}[b_j(p,q)-\mu]^2}{\tau^2}}, \tag{5.23}$$

where $b_j(p,q)$ is the pixel on the qth row, and the pth column of block b_j.

Stage 3: Encoding Pixels in Each Block

In this stage, each pixel in an image block is encoded into a binary bit of 0 if it is less than the mean and 1 if it is more, resulting in a block of binary pixels. For the jth image block, the encoded block c_j is:

$$c_j(p,q) = \begin{cases} 0 & b_j(p,q) < \mu_j \\ 1 & \text{otherwise} \end{cases}. \tag{5.24}$$

Let ϵ_j denote the number of pixels with a value of 1 in c_j; a pair of values U_j and L_j are determined as

$$U_j = \mu_j + \sigma_j\sqrt{\frac{\tau^2-\epsilon_j}{\epsilon_j}}, \tag{5.25a}$$

$$L_j = \mu_j - \sigma_j\sqrt{\frac{\epsilon_j}{\tau^2-\epsilon_j}}. \tag{5.25b}$$

The compressed data set of each image block is composed of the binary image block c_j and the pair of values U_j and L_j. Suppose U_j and L_j are each represented by r_j bits, the compressed image block contains $(2r_j+\tau^2)$ bits, and the average bit-per-pixel (bpp) is $\frac{(2r_j+\tau^2)}{\tau^2}$. The compression ratio of each image block, which is equivalent to that of the entire image, is:

$$CR = s \times \frac{\tau^2}{(2r_j+\tau^2)}, \tag{5.26}$$

where s is the number of bits for representing a pixel in the source image. For example, given $s=8$, $r_j=8$, and $\tau=4$, the bpp is $\frac{(2\times 8+4^2)}{4^2}=2$, and the compression ratio is $CR = 8\times\frac{4^2}{(2\times 8+4^2)} = 4$ times.

Decompression in BTC is rather straightforward, and conducted in a block-by-block manner. Each encrypted image block b_j' is reconstructed from its compressed data as

Figure 5.14 Images in Figures 5.13(a) and 5.13(b) that have been compressed by BTC with a compression ratio of four times.

$$b_j'(p,q) = \begin{cases} L_j & c_j(p,q) = 0 \\ U_j & \text{otherwise} \end{cases}. \tag{5.27}$$

In general, BTC is capable of compressing images with good fidelity at a moderate compression ratio. Although the compression ratio is not as high as other sophisticated methods, such as JPEG and vector quantization, BTC has the advantage of being computationally efficient, especially for the decoding process, which can be accomplished with a very small amount of arithmetic and logical operations.

To illustrate the performance of BTC, it is applied to compress the image in Figure 5.13(a) and Figure 5.13(b) with $s = 8$, $r_j = 8$, and $\tau = 4$. The decoded images, as shown in Figure 5.14, are almost identical to the original images. According to Eq. (5.26), the compression is four times, with each pixel represented by 2 bits.

An example of embedding a BTC-encoded image into a binary hologram was demonstrated in [29].

5.4.2.2 Error Diffusion

Although BTC is an effective method of reducing the data size of the embedded image, the number of bits per pixel is always larger than one. This restriction can be overcome if the image can be compressed to a 1-bit (i.e., binary) representation. A viable method to achieve this is through error diffusion, which is popularly employed in the printing industry for converting a gray level image into black and white dots. Error diffusion was first proposed by Floyd and Steinberg [32], and based on this pioneering work numerous variations from the parent scheme have been developed over the years. The method is outlined as follows. We start with an image $I(m,n)$ with pixel values within the range $[0,1]$, which is to be converted into a binary image $I_B(m,n)$. The pixel in the original image is scanned in a row-

Figure 5.15 Images from Figures 5.13 that have been converted into binary images with error diffusion.

by-row, left-to-right manner. If the value of the current pixel, located at (m_i, n_j), is larger than or equal to 0.5, it will be set to 1, otherwise it is set to 0. Mathematically:

$$I_B(m_i, n_j) = \begin{cases} 1 & I(m_i, n_j) \geq 0.5 \\ 0 & \text{otherwise} \end{cases}. \tag{5.28}$$

The error resulting from forcing the current pixel to either 1 or 0 is:

$$\text{err}(m_i, n_j) = I(m_i, n_j) - I_B(m_i, n_j). \tag{5.29}$$

Subsequently, the values of the four pixels adjacent to the current one are updated as

$$I(m_{i+1}, n_j) \leftarrow I(m_{i+1}, n_j) + w_1 \text{err}(m_i, n_j), \tag{5.30a}$$

$$I(m_{i-1}, n_{j+1}) \leftarrow I(m_{i-1}, n_{j+1}) + w_2 \text{err}(m_i, n_j), \tag{5.30b}$$

$$I(m_i, n_{j+1}) \leftarrow I(m_i, n_{j+1}) + w_3 \text{err}(m_i, n_j), \tag{5.30c}$$

$$I(m_{i+1}, n_{j+1}) \leftarrow I(m_{i+1}, n_{j+1}) + w_4 \text{err}(m_i, n_j). \tag{5.30d}$$

The constants $[w_1, w_2, w_3, w_4]$ are known as the diffusion parameters, and are assigned values of $\left[\frac{7}{16}, \frac{5}{16}, \frac{3}{16}, \frac{1}{16}\right]$ in the Floyd–Steinberg algorithm. After all the pixels have been visited, $I_B(m, n)$ will become a binary representation of the source image $I_B(m, n)$.

To illustrate the performance of error diffusion, it is applied to convert the images in Figure 5.13 into a pair of binary images, as shown in Figure 5.15. Although the source images have been reduced to a 1-bit representation, they are perceptually similar to the original images.

An attempt to embed a binary image into a binary hologram was demonstrated in [33]. In this method, the gray level image to be embedded is converted into a 1-bit binary image with error diffusion. Next, a binary hologram is generated with the downsampling technique [32]. The hologram pixels are then randomly selected and replaced by the pixels of the binary image, resulting in an intensity image-embedded binary hologram. Experimental results reveal that the binary hologram is capable of preserving both the reconstructed and the embedded images in good quality.

The same principle can be applied to the DEED hologram. The image to be embedded is first converted into a 1-bit binary image. As each hologram pixel is turned into a phase-only quantity by discarding the magnitude component, a binary pixel of the image is embedded to the least-significant-bit of the hologram pixel. The error caused by the modification of the hologram pixel is then distributed to its four immediate neighbors.

5.5 MATLAB Simulation

This section includes the MATLAB source codes that are employed to conduct the simulations of the methods reported in this chapter. A summary of the simulation programs (each identified by the name of its main function call) and their correspondence to previous sections are listed in Table 5.2.

The images listed in Table 5.3 are used in the above simulations.

5.5.1 Simulation of the SRPE Method

The simulation program is shown in Table 5.4. The main function "SRPE" loads the source image "Fish_1," and a larger image "Canvas" that comprises a QR code and two lines of text messages. The pair of images are merged into a global image. Next, a complex-valued digital Fresnel hologram is generated from the global image with the function "Holo_gen." The function "UERD" is applied to convert the hologram into a phase-only hologram with uni-directional error diffusion. An SRPE hologram is obtained by adding random phase noise to all the pixels in the phase-only hologram.

Table 5.2 List of MATLAB functions for simulating selected methods in this chapter

Function name	Description of simulation	Associated sections
SRPE	Encrypting a phase-only hologram with the single random phase method	5.3.3
ESRPE	Encrypting a phase-only hologram with the enhanced single random phase method	5.3.4
MPOH	Generated a phase-only hologram for representing multiple images through Arnold transform	5.3.5
DEED	Generating a data embedded phase-only hologram	5.4.1

Table 5.3 List of images used in the simulation programs

Image	Size	Simulation
Fish_1	1024 × 1024	SRPE
Canvas	2048 × 2048	SRPE
Fish_2	512 × 512	ESRPE
Flower_1	512 × 512	ESRPE
mpic1	512 × 512	MPOH
mpic2	512 × 512	MPOH
mpic3	512 × 512	MPOH
mpic4	512 × 512	MPOH
Flower_2	1024 × 1024	DEED
Grey	2048 × 2048	DEED

Table 5.4 MATLAB source code for simulating the generation of the SRPE hologram

```
function SRPE()
  close all
  clc
  warning off
  IMG1=imread('Fish_1.jpg');              %Read source image "Fish1"
  IMG2=imread('Canvas.jpg');              %Read global image "Global"
  [m n]=size(IMG1(:,:));
 sy=768;sx=120; %Translate source image
  IMG2(sy:sy+m-1,sx:sx+n-1)=IMG1;        %Embed translated image to global image
  %Show and save global image
  figure,imshow(IMG2);
  title('Global image');
  imwrite(IMG2,'SRPE_global.jpg');
  %Optical setting
  H_DIM=2048;                             %Dimension of hologram
  lamda=520e-9;                           %Wavelength of light
  psize=6.2e-6;                           %Pixel size
  z=0.3;                                  %Distance between image and hologram
                                            %plane
 %Generate complex-valued Fresnel hologram
  HG=Holo_gen(IMG2,z,psize,lamda);       %Generate hologram
  [ph, ph_srpe]=UERD(HG);
  ph_srpe=mat2gray(ph_srpe);
  figure,imshow(ph_srpe);
 title('SRPE hologram of global image');
  imwrite(ph_srpe,'SRPE_POH.jpg');
 %Reconstruct decoded SRPE hologram (i.e., noise removed)
  H=cos(ph)+j*sin(ph);
 RES=mat2gray(abs(Holo_recon(H,z,psize,lamda)));
  figure,imshow(RES);                     %Show reconstructed image
 title('Reconstructed image of decoded SRPE hologram');
 imwrite(RES,'RES_SRPE.jpg');
```

Table 5.4 (cont.)

```
end
function [ph, ph_noise]=UERD(HG)
  [N M]=size(HG(:,:));
  H_DIM=M;
  C=HG/max(HG(:));
 %Extract real and imaginary parts of the complex-valued hologram
  x=real(C);
  y=imag(C);
  ph=zeros(H_DIM,H_DIM);                %Phase-only hologram
 %================================================================
 %Convert complex-valued hologram to POH with error diffusion
 %================================================================
 %Define Floyd-Steinberg error diffusion coefficients
  w1=7/16;
  w2=3/16;
  w3=5/16;
  w4=1/16;
  for i = 1:H_DIM-1                     %Visit hologram pixels row by row
   for j = 2:H_DIM-1                    %From left to right in odd row
    x0=x(i,j);                          %Get real part of current pixel
    y0=y(i,j);                          %Get imaginary part of current pixel
    ang=angle(x0+1i*y0);                %Compute angle of current pixel
    ph(i,j)=ang;
   %Force magnitude of current pixel to unity
    x(i,j)=cos(ang);
    y(i,j)=sin(ang);
   %Compute error of converting complex to phase-only value
    err_x = x0-x(i,j);                  %Error of real part
    err_y = y0-y(i,j);                  %Error of imaginary part
   %Diffuse error of real part to the four neighboring pixels
    x(i,j+1)=x(i,j+1)+ w1*err_x;
   x(i+1,j-1)=x(i+1,j-1)+ w2*err_x;
    x(i+1,j)=x(i+1,j)+ w3*err_x;
   x(i+1,j+1)=x(i+1,j+1)+ w4*err_x;
   %Diffuse error of imaginary part to the four neighboring pixels
    y(i,j+1)=y(i,j+1)+ w1*err_y;
   y(i+1,j-1)=y(i+1,j-1)+ w2*err_y;
    y(i+1,j)=y(i+1,j)+ w3*err_y;
   y(i+1,j+1)=y(i+1,j+1)+ w4*err_y;
   end
  end
 %Fill in boundary pixels with neighboring pixels
  for i = 1:H_DIM
   ph(i,1)=ph(i,2);
   ph(i,H_DIM)=ph(i,H_DIM-1);
  end
  for i = 1:H_DIM
   ph(H_DIM,i)=ph(H_DIM-1,i);
   ph(1,i)=ph(2,i);
```

Table 5.4 (cont.)

```
  end
   RN=rand([H_DIM,H_DIM])*2*pi;
   two_pi=2*pi;
   ph_noise=mod(ph+RN,two_pi);
end
function H=Holo_gen(I,z,psize,lamda)
   [N M]=size(I(:,:));
   wn=2*pi/lamda;                       %Wave number
   %Generate Fresnel zone plate
   z2=z*z;                              %Square of the depth value z
  [kx ky]=meshgrid(-M/2:M/2-1,-N/2:N/2-1);
   kx=kx*psize;                         %Physical horizontal position
   ky=ky*psize;                         %Physical vertical position
  fzp=exp(1i*wn*sqrt(kx.^2+ky.^2+z2));
  %Compute hologram of source image
   A=fft2(I);                           %Fourier transform of the hologram
   B=fft2(fzp);                         %Fourier transform of the free space
                                          %impulse response
   H=fftshift(ifft2(A.*B));
end
function RES=Holo_recon(H,z,psize,lamda)
[N M]=size(H(:,:));                     %Size of hologram
wn=2*pi/lamda;                          %Wave number
%Generate Fresnel zone plate
z2=z*z;                                 %Square of the depth value z
[kx ky]=meshgrid(-M/2:M/2-1,-N/2:N/2-1);
kx=kx*psize;                            %Physical horizontal position
ky=ky*psize;                            %Physical vertical position
r=sqrt(kx.^2+ky.^2+z2);
fzp_c=exp(-1i*wn*r)./r;                 %Conjugate of FZP
A=fft2(H);                              %Fourier transform of the hologram
B=fft2(fzp_c);                          %Fourier transform of the free space
                                          %impulse response
RES=fftshift(ifft2(A.*B));              %Compute reconstructed image
End
```

Subsequently the reconstructed image of the decoded SRPE hologram at the focused plane is derived with the function "Holo_recon." The SRPE hologram and its reconstructed image are displayed and saved.

5.5.2 Simulation on the ESRPE Method

The source code for the simulation is shown in Table 5.5. Upon execution of the main function "ESRPE," an inline hologram of the source image "Fish_2" and an off-axis hologram of the secondary image "Flower_1" are generated, and summed up as a global

Table 5.5 Source code for simulating the ESRPE method

```
function ESRPE()
  %Read source image and store in 2-D array IMG1
  IMG1=imread('Fish_2.jpg');
  IMG1=im2double(IMG1);
  %Read secondary image and store in 2-D array IMG2
  IMG2=imread('Flower_1.jpg');          %Embedded image in 8-bit
                                          %representation
  IMG2=im2double(IMG1);
  %Dimension of image, assume both images have same size
  [n m]=size(IMG1(:,:));
  %Optical setting
  H_DIM=1280;                           %Dimension of hologram
  lamda=520e-9;                         %Wavelength of light
  psize=6.4e-6;                         %Pixel size
  z=0.3;                                %Distance between image and hologram
                                          %plane
  theta=0.8;                            %Off-axis angle
  %Place each input image into a larger image that is identical in size to the
  %hologram
  I1=zeros(H_DIM,H_DIM);
  I2=I1;
  I1((H_DIM-n)/2:(H_DIM+n)/2-1,(H_DIM-m)/2:(H_DIM+m)/2-1)=IMG1;
  I2((H_DIM-n)/2:(H_DIM+n)/2-1,(H_DIM-m)/2:(H_DIM+m)/2-1)=IMG2;
  RN=rand([H_DIM,H_DIM])*2*pi;          %Random phase noise normalized to
                                          %[0,255]
  %Generate holograms of the source and the secondary images
  H1=Holo_gen(I1,z,psize,lamda);        %Source hologram
  H2=Holo_gen(I2,z,psize,lamda);        %Secondary hologram
  %Convert the hologram of the secondary image to an off-axis hologram
  H2=Off_axis(H2,psize,lamda,theta);
  %Add source and secondary holograms to an integrated hologram
  HG=H1+H2;
  %Convert integrated hologram to an integrated phase-only hologram with error
  %diffusion
  ph=UERD(HG);
  %Add random phase mask to phase-only hologram of global image to give an ESRPE
  %hologram
  two_pi=2*pi;
  ph_esrpe=mod(ph+RN,two_pi);           %ESRPE phase-only hologram
  %ESRPE hologram in complex-valued representation
  H_esrpe=cos(ph_esrpe)+j*sin(ph_esrpe);
  %Decoded ESRPE hologram in complex-valued representation
  %Note decoded hologram is same as the global hologram
  HP=cos(ph)+j*sin(ph);
  ph=mat2gray(ph);
  ph_esrpe=mat2gray(ph_esrpe);
  figure,imshow(ph_esrpe);
  title('ESRPE hologram');
  imwrite(ph_esrpe,'ESRPE_POH.jpg');
  figure,imshow(ph);title('Global phase-only hologram of integrated images');
```

Table 5.5 (cont.)

```
  imwrite(ph,'ESRPE_ORG_POH.jpg');
  %Compute and display reconstructed images of ESRPE and global hologram
  RES_esrpe=mat2gray(abs(Holo_recon(H_esrpe,z,psize,lamda)));
  figure,imshow(RES_esrpe);
  title('Reconstructed image of ESRPE hologram');
  RES=mat2gray(abs(Holo_recon(HP,z,psize,lamda)));
  figure,imshow(RES);
  title('Reconstructed image of original/decoded global hologram');
  imwrite(RES,'RES_ESRPE.jpg');
end
function H=Off_axis(H,psize,lamda,
   theta)
  theta=theta*pi/180;
  [N M]=size(H(:,:))
  [kx ky]=meshgrid(-N/2:N/2-1,-M/2:M/2-1);
  R=exp(1i*2*pi.*kx*psize*sin(theta)/lamda);
  H=H.*R;
end
function ph=UERD(HG)
  [N M]=size(HG(:,:));
  H_DIM=M;                              %Assume a square hologram
  C=HG/max(HG(:));
  %Extract real and imaginary parts of the complex-valued hologram
  x=real(C);
  y=imag(C);
  ph=zeros(H_DIM,H_DIM);                %Declare phase-only hologram
  %=====================================================
  %Convert complex-valued hologram to POH with error diffusion
  %======================================================
  %Define Floyd-Steinberg error diffusion coefficients
  w1=7/16;
  w2=3/16;
  w3=5/16;
  w4=1/16;
  for i = 1:H_DIM-1                     %Visit hologram pixels row by row
   for j = 2:H_DIM-1                    %From left to right in odd row
     x0=x(i,j);                         %Get real part of current pixel
     y0=y(i,j);                         %Get imaginary part of current pixel
     ang=angle(x0+1i*y0);               %Compute angle of current pixel
     ph(i,j)=ang;
     %Force magnitude of current pixel to unity
     x(i,j)=cos(ang);
     y(i,j)=sin(ang);
     %Compute error of converting complex to phase-only value
     err_x = x0-x(i,j);                 %Error of real part
     err_y = y0-y(i,j);                 %Error of imaginary part
     %Diffuse error of real part to the four neighboring pixels
     x(i,j+1)=x(i,j+1)+ w1*err_x;
     x(i+1,j-1)=x(i+1,j-1)+ w2*err_x;
     x(i+1,j)=x(i+1,j)+ w3*err_x;
```

Table 5.5 (cont.)

```
      x(i+1,j+1)=x(i+1,j+1)+ w4*err_x;
      %Diffuse error of imaginary part to the four neighboring pixels
      y(i,j+1)=y(i,j+1)+ w1*err_y;
      y(i+1,j-1)=y(i+1,j-1)+ w2*err_y;
      y(i+1,j)=y(i+1,j)+ w3*err_y;
      y(i+1,j+1)=y(i+1,j+1)+ w4*err_y;
    end
  end
  %Fill in boundary pixels with neighboring pixels
  for i = 1:H_DIM
    ph(i,1)=ph(i,2);
    ph(i,H_DIM)=ph(i,H_DIM-1);
  end
  for i = 1:H_DIM
    ph(H_DIM,i)=ph(H_DIM-1,i);
    ph(1,i)=ph(2,i);
  end
  end
function H=Holo_gen(I,z,psize,lamda)
  [N M]=size(I(:,:));
  wn=2*pi/lamda;                          %Wave number
  %Generate Fresnel zone plate
  z2=z*z;                                 %Square of the depth value z
  [kx ky]=meshgrid(-M/2:M/2-1,-N/2:N/2-1);
  kx=kx*psize;                            %Physical horizontal position
  ky=ky*psize;                            %Physical vertical position
  fzp=exp(1i*wn*sqrt(kx.^2+ky.^2
  +z2));
  %Compute hologram of source image
  A=fft2(I);                              %Fourier transform of the hologram
  B=fft2(fzp);                            %Fourier transform of the free space
                                            %impulse response
  H=fftshift(ifft2(A.*B));
end
function RES=Holo_recon(H,z,psize,lamda)
  [N M]=size(H(:,:)); %Size of hologram
  wn=2*pi/lamda;                          %Wave number
  %Generate Fresnel zone plate
  z2=z*z;                                 %Square of the depth value z
  [kx ky]=meshgrid(-M/2:M/2-1,-N/2:N/2-1);
  kx=kx*psize;                            %Physical horizontal position
  ky=ky*psize;                            %Physical vertical position
  r=sqrt(kx.^2+ky.^2+z2);
  fzp_c=exp(-1i*wn*r)./r;                 %Conjugate of FZP
  A=fft2(H);                              %Fourier transform of the hologram
  B=fft2(fzp_c);                          %Fourier transform of the free space
                                            %impulse response
  RES=fftshift(ifft2(A.*B));              %Compute reconstructed image
end
```

hologram. Next, the uni-directional error diffusion function "UERD" is applied to convert the global hologram into a phase-only hologram. An ESRPE hologram is obtained by adding random phase noise to all the pixels in the phase-only hologram.

Subsequently, the reconstructed image of the decoded ESRPE hologram at the focused plane is derived with the function "Holo_recon." The ESRPE hologram and the reconstructed image of the decoded ESRPE hologram are then displayed and saved. In the reconstructed image, the source is located at the center, while the secondary image is displaced and separated from the source image.

5.5.3 Simulation of Multiple-Image Holographic Encryption with Arnold Transform

In this simulation, four source images, "mpic1," "mpic2," "mpic3," and "mpic4," are encoded into a single phase-only hologram. The simulation program is shown in Table 5.6.

Table 5.6 Simulation program on the MPOH method

```
function MPOH()
  close all
  clc
  warning off
  %Optical settings and downsampling factor
  H_DIM=1024;                          %Dimension of hologram
  lamda=520e-9;                        %Wavelength of light
  dx=6.4e-6;                           %Pixel size
  dist=0.3;                            %Distance between image and hologram
                                         %plane
  DS=12;
  %Load the four source images
  I1=Add_image('mpic1.jpg',DS);
  I2=Add_image('mpic2.jpg',DS);
  I3=Add_image('mpic3.jpg',DS);
  I4=Add_image('mpic4.jpg',DS);
  [row col]=size(I1(:,:));
  IMG1=zeros(H_DIM,H_DIM);
  IMG2=IMG1;IMG3=IMG1;IMG4=IMG1;
  sx=(H_DIM-col)/2;ex=(H_DIM+col)/2-1;
  sy=(H_DIM-row)/2;ey=(H_DIM+row)/2-1;
  IMG1(sy:ey,sx:ex)=I1;
  IMG2(sy:ey,sx:ex)=I2;
  IMG3(sy:ey,sx:ex)=I3;
  IMG4(sy:ey,sx:ex)=I4;
  %Generate POH with the CAM method, and return reconstructed image
  H1=Holo_gen(IMG1,dist,dx,lamda);     %Generate first hologram
  H=Arnold(H1,H_DIM,10);
```

Table 5.6 (cont.)

```
  H2=Holo_gen(IMG2,dist,dx,lamda);     %Generate second hologram
  H=H+Arnold(H2,H_DIM,20);
  H3=Holo_gen(IMG3,dist,dx,lamda);     %Generate third hologram
  H=H+Arnold(H3,H_DIM,30);
  H4=Holo_gen(IMG4,dist,dx,lamda);     %Generate fourth hologram
  H=H+Arnold(H4,H_DIM,40);
  ANG=angle(H);
  HP_EN=cos(ANG)+j*sin(ANG);
  HP=I_Arnold(HP_EN,H_DIM,10);
  RES=abs(Holo_recon(HP,dist,dx,       %Reconstruct first hologram
   lamda));
  RES=mat2gray(RES);
  figure,imshow(RES);                  %Show reconstructed image
  title('Reconstruction with 5 iterations');
  imwrite(RES,'RES_MPOH_1.jpg');
  HP=I_Arnold(HP_EN,H_DIM,20);
  RES=abs(Holo_recon(HP,dist,dx,       %Reconstruct second hologram
   lamda));
  RES=mat2gray(RES);
  figure,imshow(RES);                  %Show reconstructed image
  title('Reconstruction with 10 iterations');
  imwrite(RES,'RES_MPOH_2.jpg');
  HP=I_Arnold(HP_EN,H_DIM,30);
  RES=abs(Holo_recon(HP,dist,dx,       %Reconstruct third hologram
   lamda));
  RES=mat2gray(RES);
  figure,imshow(RES);                  %Show reconstructed image
  title('Reconstruction with 15 iterations');
  imwrite(RES,'RES_MPOH_3.jpg');
  HP=I_Arnold(HP_EN,H_DIM,40);
  RES=abs(Holo_recon(HP,dist,dx,       %Reconstruct fourth hologram
   lamda));
  RES=mat2gray(RES);
  figure,imshow(RES);                  %Show reconstructed image
  title('Reconstruction with 20 iterations');
  imwrite(RES,'RES_MPOH_4.jpg');
end
function IMG=Add_image(fname,DS)
  I=imread(fname);
  I=im2double(I);
  [row col]=size(I(:,:));
  IMG=downsample(I,DS);
end
function IMG=downsample(I,DS)
  [m n]=size(I(:,:));
  IMG=I;
  for row=1:n
   for col=1:m
    x=mod(col,DS);
```

Table 5.6 (cont.)

```
    y=mod(row,DS);
    if x~=0 && y~=0 && x~=y && x~=(DS-y)
      %IMG(row,col)=0;
      IMG(row,col)=-IMG(row,col);
     end
   end
  end
end
function HA=Arnold(H,H_DIM,iter)
  A=[1 2;1 3];                            %Arnold transform matrix
  for row=1:H_DIM
   for col=1:H_DIM
    pos_org=[row;col]-1;
    for i=1:iter
     pos_new=mod(round(A*pos_org),H_DIM);
     pos_org=pos_new;
    end
    pos_new=pos_new+1;
    HA(pos_new(1),pos_new(2))=H(row,col);
   end
  end
end
function HA=I_Arnold(H,H_DIM,iter)
  A=[3-2;-1 1];                           %Inverse Arnold transform matrix
  for row=1:H_DIM
   for col=1:H_DIM
    pos_org=[row;col]-1;
    for i=1:iter
     pos_new=mod(round(A*pos_org),H_DIM);
     pos_org=pos_new;
    end
    pos_new=pos_new+1;
    HA(pos_new(1),pos_new(2))=H(row,col);
   end
  end
end
function H=Holo_gen(I,z,psize,lamda)
  [N M]=size(I(:,:));
  wn=2*pi/lamda;                          %Wave number
  %Generate Fresnel zone plate
  z2=z*z;                                 %Square of the depth value z
  [kx ky]=meshgrid(-M/2:M/2-1,-N/2:N/2-1);
  kx=kx*psize;                            %Physical horizontal position
  ky=ky*psize;                            %Physical vertical position
  fzp=exp(1i*wn*sqrt(kx.^2+ky.^2
  +z2));
  %Compute hologram of source image
  A=fft2(I);                              %Fourier transform of the hologram
```

Table 5.6 (cont.)

```
  B=fft2(fzp);                          %Fourier transform of the free space
                                          %impulse response
  H=fftshift(ifft2(A.*B));
end
function IR=Holo_recon(H,z,psize,lamda)
  [N M]=size(H(:,:));                   %Size of hologram
  wn=2*pi/lamda;                        %Wave number
  %Generate Fresnel zone plate
  z2=z*z;                               %Square of the depth value z
  [kx ky]=meshgrid(-M/2:M/2-1,-N/2:N/2-1);
  kx=kx*psize;                          %Physical horizontal position
  ky=ky*psize;                          %Physical vertical position
  r=sqrt(kx.^2+ky.^2+z2);
  fzp_c=exp(-1i*wn*r)./r;               %Conjugate of FZP
  A=fft2(H);                            %Fourier transform of the hologram
  B=fft2(fzp_c);                        %Fourier transform of the free space
                                          %impulse response
  IR=fftshift(ifft2(A.*B));             %Compute reconstructed image
end
```

Each of the source images is first downsampled with the function "downsample," based on a grid-cross lattice with a downsampling factor of $\tau = 12$. The function "Holo_gen" is applied to generate the Fresnel holograms of the downsampled images. Each hologram is then scrambled by a unique number of rounds of Arnold transforms with the function "Arnold." The scrambled holograms are summed up into an integrated hologram, and the phase component is extracted to be the integrated phase-only hologram.

The function "I_Arnold" is used to decode a given hologram. First the integrated phase-only hologram is inverse Arnold transformed by the corresponding number of iterations that has been imposed on the hologram in the encoding process. Subsequently, the reconstructed image of the decoded hologram at the focused plane is obtained with the function "Holo_recon," and the result is displayed and saved.

5.5.4 Simulation of Generating a DEED Hologram that Embeds an Image

In the simulation shown in Table 5.7, a DEED phase-only hologram that is embedded with an image is generated. The program is divided into two major parts. The first part generates a Fresnel hologram $H(m,n)$, which represents a source image $I(m,n)$ that is parallel to, and at an axial distance of 0.3 m from, the hologram plane. In the second part of the program, uni-directional error diffusion is applied to convert the hologram into a phase-only hologram, and in due course embedding it with a 2-D image that is identical in size to the hologram.

Table 5.7 Simulation program on the DEED method

```
function DEED()
close all;
warning off
 %Read source image and store in a 2-D array
 IMG0=imread('Flower_2.jpg');
 IMG0=im2double(IMG0);
 [n m]=size(IMG0(:,:));                 %Size of source image
 IMG1=imread('Grey.jpg');
 IMG1=im2double(IMG1)*255;
 IMG1=uint8(IMG1);
 [N M]=size(IMG1(:,:));                 %Size of embedded image
 H_DIM=M;                               %Assume hologram of size identical to
                                          %embedded image
 bL=input('Enter number of bits for embedding external data ');
 i_bit=8-bL;                            %Number of bits to represent hologram
 IMG2=bitshift(IMG1,-i_bit);
 IMG2=bitshift(IMG2,i_bit);
 figure;
 imshow(mat2gray(IMG2));                %Show de-quantized embedded image
 fname=strcat('IEED_embed_bL',num2str(bL),'.jpg');
 imwrite(mat2gray(IMG2),fname);
 %Optical setting
 lamda=520e-9;                          %Wavelength of light
 psize=6.4e-6;                          %Pixel size
 z=0.3;                                 %Distance between image and hologram
                                          %plane
 %Move source image to a larger canvas that is similar in size to the hologram
 IMG=zeros(H_DIM,H_DIM);
 IMG((H_DIM-n)/2:(H_DIM+n)/2-1,(H_DIM-m)/2:(H_DIM+m)/2-1)=IMG0;
 %Generate POH with the CAM method, and return reconstructed image
 H=Holo_gen(IMG,z,psize,lamda);         %Generate hologram
 %Convert hologram to UERD phase-only hologram
 Hp=UERD_DEED(H,IMG2,H_DIM,bL);
 RES=abs(Holo_recon(Hp,z,psize,         %Reconstruct DEED hologram
  lamda));
 %Crop the area containing the image
 RES0=RES((H_DIM-n)/2:(H_DIM+n)/2-1,(H_DIM-m)/2:(H_DIM+m)/2-1);
 figure,imshow(mat2gray(RES0));         %Show reconstructed image
 fname=strcat('RES_IEED_bL',num2str(bL),'.jpg');
 imwrite(mat2gray(RES0),fname);
end
function H=Holo_gen(I,z,psize,lamda)
 [N M]=size(I(:,:));
 wn=2*pi/lamda;                         %Wave number
 %Generate Fresnel zone plate
 z2=z*z;                                %Square of the depth value z
 [kx ky]=meshgrid(-M/2:M/2-1,-N/2:N/2-1);
 kx=kx*psize;                           %Physical horizontal position
```

Table 5.7 (cont.)

```
 ky=ky*psize;                          %Physical vertical position
 fzp=exp(1i*wn*sqrt(kx.^2+ky.^2
  +z2));
 %Compute hologram of source image
 A=fft2(I);                            %Fourier transform of the hologram
 B=fft2(fzp);                          %Fourier transform of the free space
                                         %impulse response
 H=fftshift(ifft2(A.*B));
end
function RES=Holo_recon(H,z,psize,lamda)
 [N M]=size(H(:,:));                   %Size of hologram
 wn=2*pi/lamda;                        %Wave number
 %Generate Fresnel zone plate
 z2=z*z;                               %Square of the depth value z
 [kx ky]=meshgrid(-M/2:M/2-1,-N/2:N/2-1);
 kx=kx*psize;                          %Physical horizontal position
 ky=ky*psize;                          %Physical vertical position
 r=sqrt(kx.^2+ky.^2+z2);
 fzp_c=exp(-1i*wn*r)./r;               %Conjugate of FZP
 A=fft2(H);                            %Fourier transform of the hologram
 B=fft2(fzp_c);                        %Fourier transform of the free space
                                         %impulse response
 RES=fftshift(ifft2(A.*B));            %Compute reconstructed image
end
function HOLO=UERD_DEED(H,IMG2,H_DIM,e_bit)
 [m n]=size(H(:,:));
  C=H./max(abs(H(:)));
 %Extract real and imaginary parts of the complex-valued hologram
 x=real(C);
 y=imag(C);
 i_bit=8-e_bit;                        %Number of bits that are discarded and
                                         %not embedded
 EMBED=bitshift(IMG2,-i_bit);          %Only most significant e_bits are
                                         %embedded
 ph=uint8(zeros(m,n));
 %=================================================================
 %Convert complex-valued hologram to POH with error diffusion
 %=================================================================
 %Define Floyd-Steinberg error diffusion coefficients
 w1=7/16;
 w2=3/16;
 w3=5/16;
 w4=1/16;
 for i = 1:H_DIM-1                     %Visit hologram pixels row by row
  for j = 2:H_DIM-1                    %From left to right in odd row
    x0=x(i,j);                         %Get real part of current pixel
    y0=y(i,j);                         %Get imaginary part of current pixel
    ang=angle(x0+1i*y0);               %Compute phase angle of current pixel
    if ang<0 ang=ang+2*pi;             %Convert phase angle to range [0,255]
```

Table 5.7 (cont.)

```
    end
    phase=uint8((ang/2/pi)*255);
    phase=bitshift(phase,-e_bit);
    phase=bitshift(phase,e_bit);
    ph(i,j)=bitor(phase,EMBED(i,j));
    %Replace the least e_bits with the embedded image
    ang=2*pi*double(ph(i,j))/255;
    %Force magnitude of current pixel to unity
    x(i,j)=cos(ang);
    y(i,j)=sin(ang);
    %Compute error of converting complex to phase-only value
    err_x = x0-x(i,j);                    %Error of real part
    err_y = y0-y(i,j);                    %Error of imaginary part
    %Diffuse error of real part to the four neighboring pixels
    x(i,j+1)=x(i,j+1)+ w1*err_x;
    x(i+1,j-1)=x(i+1,j-1)+ w2*err_x;
    x(i+1,j)=x(i+1,j)+ w3*err_x;
    x(i+1,j+1)=x(i+1,j+1)+ w4*err_x;
    %Diffuse error of imaginary part to the four neighboring pixels
    y(i,j+1)=y(i,j+1)+ w1*err_y;
    y(i+1,j-1)=y(i+1,j-1)+ w2*err_y;
    y(i+1,j)=y(i+1,j)+ w3*err_y;
    y(i+1,j+1)=y(i+1,j+1)+ w4*err_y;
   end
  end
  for i = 1:H_DIM
   x(i,1)=x(i,2);
   x(i,H_DIM)=x(i,H_DIM-1);
   ph(i,1)=ph(i,2);
   ph(i,H_DIM)=ph(i,H_DIM-1);
  end
  for i = 1:H_DIM
   x(H_DIM,i)=x(H_DIM-1,i);
   x(1,i)=x(2,i);
   ph(H_DIM,i)=ph(H_DIM-1,i);
   ph(1,i)=ph(2,i);
  end
  %=================================================================
  %Store the phase-only hologram to the output array 'HOLO'
  %=================================================================
  r=(double(ph)/255)*2*pi;
  HOLO=cos(r)+1i*sin(r);
End
```

When the program is executed, the user will be prompted to enter the number of bits in the DEED hologram to be used for embedding the external image. Upon completion of the process, the DEED hologram and the embedded image will be displayed and saved.

5.6 Summary

In this chapter, a number of practical applications of phase-only holograms have been described. In Section 5.2 the projection and display of phase-only holograms on contemporary SLM devices was outlined. The optical setup required for showing a phase-only hologram is simplified significantly. Section 5.3 explains the art of optical/holographic encryption. The classical double random phase encryption (DRPE), which has been widely adopted in optical cryptography, was introduced. Two recent techniques, single random phase encryption and enhanced single random phase encryption, were sequentially presented. Both methods are capable of encrypting a phase-only hologram with high resistance to the common KPAs. After that, Arnold transform for encrypting multiple images in a single phase-only hologram was described. Finally, in Section 5.4, the art of embedding external data onto a phase-only hologram, a technique commonly known as steganography was demonstrated. First, the DEED method, which incorporates error diffusion for converting a complex-valued hologram into a phase-only hologram, and in due course embedding external data into each hologram pixel, was elucidated. Evaluation shows that the DEED method is effective, but the number of bits that can be used for embedding external data has to be limited to a small bit length in order to preserve the fidelity of a hologram and its reconstructed image. However, if the embedded data is an image, then it has to be truncated to fewer bits, leading to heavy distortion of the image. This dilemma is overcome with the image embedded error diffusion (IEED) method. The principle of IEED is to embed a compressed image – instead of the original image – into a DEED hologram. Two simple techniques, BTC and error diffusion, were implemented to convert a gray level image into a 2-bit and a 1-bit representation, meaning it could be embedded into a DEED hologram without too much distortion. The results also show that despite the reduction in bit length, the quality of the image compressed with BTC or error diffusion is well preserved.

Exercises

E.5.1 What are the disadvantages of the off-axis holographic projection in Figure 5.2(b)?

E.5.2 Write a MATLAB program to show how the encryption key of the DRPE method in Figure 5.4(a) can be deduced with a Dirac delta function as the source image.

E.5.3 Based on the MATLAB simulation program in Section 5.5.2, evaluate the effect of the angles of incidence $(\theta_{k;m}, \theta_{k;n})$ on the reconstructed images (for both the source and the secondary images) of the decrypted hologram. From the result, explain what should be the appropriate angles of incidence for these images.

E.5.4 A 4×4 image block with pixel intensity in the range $[0, 255]$ is given below. Determine the pixel values in the image block after a single round of Arnold transform.

E.5.5 In Section 5.3.5, grid-cross downsampling (GCD) is applied to the source images before they are converted and integrated into a single hologram via Arnold transform. As a result, only the phase component of the hologram is needed to represent the multiple images. The phase-only hologram generated with this method is referred to as an integrated sampled phase-only hologram (ISPOH). Based on the MATLAB simulation program in Section 5.5.3, explore other methods, such as random phase noise addition, for generating an integrated phase-only hologram for representing multiple images.

E.5.6 In Section 5.4.1, simulation on the data embedded error diffusion (DEED) hologram is conducted with embedded bit lengths of $b_L = 2$ and 6, based on the MATLAB code in Section 5.5.4. Extend the evaluation with other values of b_L, and determine which one gives a good compromise between the quality of the embedded image and the reconstructed image of the DEED hologram.

E.5.7 Block truncation coding is applied to compress the image block in E.5.4 based on a block size of 4×4. Determine and show the compressed data.

E.5.8 A pixel P in an image has an intensity value of 0.32, and is binarized with Floyd–Steinberg error diffusion. The dynamic range of the intensity of the image is [0,1]. Determine the error that is distributed to the four neighboring pixels of P.

References

[1] P. Refregier and B. Javidi, "Optical image encryption based on input plane and Fourier plane random encoding," *Opt. Lett.*, vol. 20, pp. 767–769 (1995).

[2] S. Jiao, G. Li, C. Zhou, W. Zou, and X. Li, "Special ciphertext-only attack to double random phase encryption by plaintext shifting with speckle correlation," *J. Opt. Soc. Am. A*, vol. 35, pp. A1–A6 (2018).

[3] G. Li, W. Yang, D. Li, and G. Situ, "Cyphertext-only attack on the double random-phase encryption: experimental demonstration," *Opt. Exp.*, vol. 25, pp. 8690–8697 (2017).

[4] X. Liu, J. Wu, W. He, M. Liao, C. Zhang, and X. Peng, "Vulnerability to ciphertext-only attack of optical encryption scheme based on double random phase encoding," *Opt. Exp.*, vol. 23, pp. 18955–18968 (2015).

[5] Y. Zhang, D. Xiao, W. Wen, and H. Liu, "Vulnerability to chosen-plaintext attack of a general optical encryption model with the architecture of scrambling-then-double random phase encoding," *Opt. Lett.*, vol. 38, pp. 4506–4509 (2013).

[6] Y. Frauel, A. Castro, T. Naughton, and B. Javidi, "Resistance of the double random phase encryption against various attacks," *Opt. Exp.*, vol. 15, pp. 10253–10265 (2007).

[7] X. Peng, H. Wei, and P. Zhang, "Chosen-plaintext attack on lensless double-random phase encoding in the Fresnel domain," *Opt. Lett.*, vol. 31, pp. 3261–3263 (2006).

[8] X. Peng, P. Zhang, H. Wei, and B. Yu, "Known-plaintext attack on optical encryption based on double random phase keys," *Opt. Lett.*, vol. 31, pp. 1044–1046 (2006).

[9] A. Carnicer, M. Montes-Usategui, S. Arcos, and I. Juvells, "Vulnerability to chosen-cyphertext attacks of optical encryption schemes based on double random phase keys," *Opt. Lett.*, vol. 30, pp. 1644–1646 (2005).

[10] W. Chen and X. Chen, "Double random phase encoding using phase reservation and compression," *J. Opt.*, vol. 16, art. no. 205402 (2014).

[11] A. Markman and B. Javidi, "Full-phase photon-counting double-random-phase encryption," *J. Opt. Soc. Am. A*, vol. 31, pp. 394–403 (2014).

[12] Q. Gong, X. Liu, G. Li, and Y. Qin, "Multiple-image encryption and authentication with sparse representation by space multiplexing," *Appl. Opt.*, vol. 52, pp. 7486–7493 (2013).

[13] G. Unnikrishnan, J. Joseph, and K. Singh, "Optical encryption by double-random phase encoding in the fractional Fourier domain," *Opt. Lett.*, vol. 25, pp. 887–889 (2000).

[14] R. Tao, Y. Xin, and Y. Wang, "Double image encryption based on random phase encoding in the fractional Fourier domain," *Opt. Exp.*, vol. 15, pp. 16067–16079 (2007).

[15] P. Kumar, A. Kumar, J. Joseph, and K. Singh, "Impulse attack free double-random-phase encryption scheme with randomized lens-phase functions," *Opt. Lett.*, vol. 34, pp. 331–333 (2009).

[16] W. Chen, C. Quan, C.J. Tay, "Optical color image encryption based on Arnold transform and interference method," *Opt Comm.*, vol. 282, no. 18, pp. 3680–3685 (2009).

[17] Z. Ren, P. Su, J. Ma, and G. Jin, "Secure and noise-free holographic encryption with a quick-response code," *Chin. Opt. Lett.*, vol. 12, art. no. 010601 (2014).

[18] J. Barrera, A. Mira, and R. Torroba, "Optical encryption and QR codes: secure and noise-free information retrieval," *Opt. Exp.*, vol. 21, pp. 5373–5378 (2013).

[19] P.W.M. Tsang, "Single-random-phase holographic encryption of images," *Opt. Lasers Eng.*, vol. 89, pp. 22–28 (2017).

[20] P.W.M. Tsang, "Enhanced single random phase holographic encryption of optical images," in *Advanced Secure Optical Image Processing for Communications*, A.A. Falou (ed.) (IOP Publishing, 2018), pp. 9-1–9-12.

[21] P.W.M. Tsang, A. Jiao, and T.-C. Poon, "Fast conversion of digital Fresnel hologram to phase-only hologram based on localized error diffusion and redistribution," *Opt. Exp.*, vol. 22, pp. 5060–5066 (2014).

[22] P.W.M. Tsang and T.-C. Poon, "Novel method for converting digital Fresnel hologram to phase-only hologram based on bidirectional error diffusion," *Opt. Exp.*, vol. 21, pp. 23680–23686 (2013).

[23] P.W.M. Tsang, T.-C. Poon, W. Wang, X. Zhu, and K. Chan, "Integrating multiple images in a sampled phase-only hologram," *Chin. Opt. Lett.*, vol. 17, art. no. 050901 (2019).

[24] B. Li and J. W. Xu, "Period of Arnold transformation and its application in image scrambling," *J. Cen. South U. Tech.*, vol. 12, p. 278 (2005).

[25] E. Hariyanto and R. Rahim, "Arnold's cat map algorithm in digital image encryption," *Intl J. Sci. Res.*, vol. 5, pp. 1363–1365 (2016).

[26] M. Mishra, S. Kumar, and S. Mishra, "Security enhanced digital image steganography based on successive Arnold transformation," in *Advances in Computer Science, Engineering & Applications*, D. Wyld, J. Zizka, and D. Nagamalai (eds.) (Springer, 2012), pp. 221–229.

[27] M. Li, T. Liang, and Y.-J. He, "Arnold transform based image scrambling method," *Proc. 3rd Intl Conf. Multimedia Technology (ICMT-13)* (2013).

[28] P.W.M. Tsang and T.-C. Poon, "Data-embedded-error-diffusion hologram (invited paper)," *Chin. Opt. Lett.*, vol. 12, art. no. 060017 (2014).

[29] P.W.M. Tsang, T.-C. Poon, and A.S.M. Jiao, "Embedding intensity image in grid-cross down-sampling (GCD) binary holograms based on block truncation coding," *Opt. Comm.*, vol. 304, p. 62 (2013).

[30] P.W.M. Tsang, T.-C. Poon, and W.K. Cheung, "Intensity image-embedded binary holograms," *Appl. Opt.*, vol. 52, pp. A26–A32 (2013).

[31] E. Delp and O. Mitchell, "Image compression using block truncation coding," *IEEE Trans. Comm.*, vol. 27, no. 9, p. 1335 (1979).

[32] R.W. Floyd and L. Steinberg, "An adaptive algorithm for spatial grey scale," *Proc. Soc. Inf. Disp.*, vol. 17, pp. 75–77 (1976).

[33] P.W.M. Tsang, T.-C. Poon, W.K. Cheung, and J.P. Liu, "Computer generation of binary Fresnel holography," *Appl. Opt.*, vol. 50, pp. B88–B95 (2011).

Index

For EU product safety concerns, contact us at Calle de José Abascal, 56–1°,
28003 Madrid, Spain or eugpsr@cambridge.org.

www.ingramcontent.com/pod-product-compliance
Lightning Source LLC
LaVergne TN
LVHW062210200926
848484LV00006B/1365